Creative Kids

Simple Science Fun

Written by Evan Forbes,
Janet Hale, and
Cindy Christianson

Teacher Created Resources

Teacher Created Resources, Inc.
6421 Industry Way
Westminster, CA 92683
www.teachercreated.com

ISBN: 978-0-7439-3196-0

2004 Teacher Created Resources, Inc.
Reprinted, 2009
Made in the U.S.A.

Editors:

Wanda Kelly

JoAnn Merrell

Cover Artist

Tony Carrillo

Illustrator

Wendy Chang

Kelly McMahon

Table of Contents

Table of Contents

About This Book

Science is the study of the physical and biological environment. Children who are exposed to the wonders of science in an enjoyable and caring way will begin to appreciate the beauty of the unique world in which we live.

Science is more than the study of facts and the acquisition of basic scientific knowledge. It involves observing, predicting, testing (experiencing trial and error), discovering, and discussing results. *Simple Science Fun* has been designed with these processes in mind.

There are three areas of science study: life (e.g., growth, food chains, and habitats), earth (e.g., air, soil, and water), and physical (e.g., matter, light, and force). *Simple Science Fun* ventures into all three of these areas. Each experience has been designed to be quick and easy. With only minor or no preparation required, children are introduced to science experiences which are amazing, exciting, and intellectually stimulating.

Why should *Simple Science Fun* be an important supplement to your child's educational experience?

Understanding the principles of science is essential to children if they are to cope successfully with our ever-changing world. Decision making is a critical skill. *Simple Science Fun* causes children to predict and make decisions based on prior knowledge, as well as newly acquired knowledge. *Simple Science Fun* helps children become aware of the concept that "what we expect to happen often times does not." The skill of making confident decisions will prove to be one of the most essential tools for children to function as happy and healthy persons.

Understanding that science has rules that remain constant, children begin to build confidence and develop a willingness to take risks. For example, no matter what lesson is being taught, the principles of inertia (an object's resistance to change in motion, depending on its mass) always perform with 100% accuracy. Insight into these science consistencies benefit children when they begin to make inferences in other subject areas.

An appreciation of science brings personal pleasure and satisfaction. Enjoyment is an excellent motivator. To be effective, the study of science should be enjoyable to everyone involved. For too many years science has been taught with only textbooks, worksheets, and very controlled experiments. This not only gets boring for children but the parent/teacher, as well. If the science experienced is fun for all, all will experience the joy of learning science.

Science Safety Rules

1. Begin science activities only after all of the steps have been reviewed.

2. There should always be an adult present when working with any kind of sharp object (e.g., scissors, craft knife, safety pins, skewers, toothpicks, etc.).

3. Never put anything in your mouth unless it is required by the science experiment and there is adult supervision.

4. Raw eggs have the potential to cause salmonella poisoning. Wash the inside and outside of egg shells with a little bleach before using.

5. When collecting plants, be sure to caution your child to collect only in permitted areas and to collect gently. They should also know how to recognize poisonous plants in your area and know not to touch the parts of a plant they are uncertain about. Warn them not to eat the parts of any plants they collect in the wild, unless they have been assured of their safety by a knowledgeable adult.

6. Never remove an item that is being used as a home for an animal. Respect all living creatures.

7. Clean up and dispose of waste and recyclables in proper containers when finished with an activity.

You can have fun and be safe at the same time!

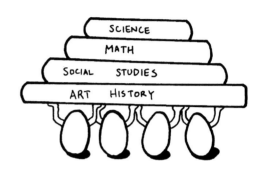

Physical Science Activities

Amazing Balloon

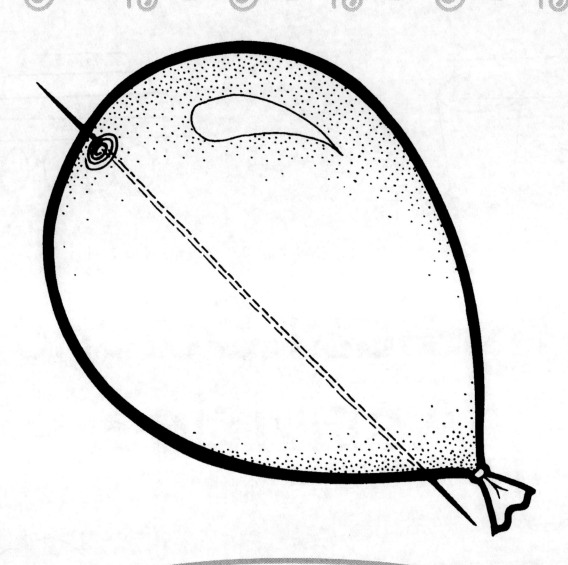

Question

Why can you run a skewer through an inflated balloon without it popping?

Amazing Balloon

Materials

- several 8–10" (20–25 cm) round-shaped balloons
- one thin, wooden shish-kebab skewer
- petroleum jelly

What to Do

1. Blow up one balloon (not too full). It works best when the balloon is blown to half-size. Then tie the end of the balloon in a knot. Stick a wooden skewer into the balloon. What happens?
2. Blow up a second balloon just like in step one. This time you are going to attempt to put the skewer through the entire balloon without popping it.
3. Coat the tip of your skewer with petroleum jelly and then push the sharp end of the skewer into the balloon right next to where the knot is tied.
4. Continue to push the skewer gently through the inside cavity of the balloon and then exit the balloon at the top of the balloon where the rubber is least stretched. (This area looks like a small, dark dot or circle.)
5. When the point of the skewer begins to exit the balloon, continue pushing so the skewer is even on both sides.
6. Why did the balloon not pop?

Why It Works

Balloons are made of rubberized material. This rubberized material is called a *polymer*. Poly- meaning many and mer- meaning molecules. The balloon's material is made up of many molecules linked together. These links are strong and try hard to stay linked (they do not like being broken or pulled apart). When the skewer is pushed through the balloon where it is least stretched (unlike the first popping experience) the links give just enough, allowing the skewer to pass through the balloon. This also creates a hole in the balloon, allowing air to slowly leak out. This slow leak becomes obvious if the skewered balloon is observed about one-half hour later.

An Underwater Fountain

Question

What causes colored water to rise from a small bottle?

An Underwater Fountain

Materials

- small glass bottle
- hot water
- food coloring of your choice
- string
- large jar of cold water

What to Do

1. Tie the string around the neck of the small bottle.

2. Then fill the bottle with hot water and add food coloring.

3. Slowly lower the bottle into the large jar of cold water. A cloud like a colored fountain will rise out of the small bottle.

Why It Works

Water expands when it is heated. Hot water is less dense than the cold water. Therefore, the hot water floats to the top just as wood or a cork would float. They are less dense, as well.

Baffling Blast of Air

Question

Is there more behind the cardboard than meets the eye?

Baffling Blast of Air

Materials

- one ruler
- one pencil
- one thumbtack
- one thread spool (thread can still be on spool)
- one 4" (10 cm) square piece of light cardboard

What to Do

1. Find the centerpoint of the cardboard and mark it with a pencil. Then push the thumbtack through the centerpoint.

2. On a flat surface, place the thread spool over the thumbtack point so the point of the thumbtack is located in the center hole of the spool. What do you think will happen when you pick up the spool and the cardboard and blow a long blast of air through the top hole of the thread spool?

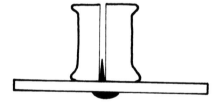

3. Why did the cardboard stay in place and not fall away?

Why It Works

When air is blown through the center hole of the spool, it causes an air stream. This air stream passes between the bottom of the spool and the cardboard. Moving air has less pushing power than the air below the cardboard and causes the card to be pushed upward towards the spool. The thumbtack is important because it causes the cardboard to stay in place rather than be pulled in the direction of the strongest point of the blowing air stream.

Balancing a Potato

Question

Do you think you can balance a potato on the edge of a glass?

Balancing a Potato

Materials

- a glass
- one small raw potato
- two identical forks

What to Do

1. Place the glass on a flat surface.

2. Push the prongs of one fork upward into one side of the potato. Repeat this with the second fork on the opposite side of the potato. Both fork handles should stick out at the same angle on each side.

3. Center the potato on the edge of the glass. Adjust its position and the forks until they balance.

4. What allows the potato to balance on the edge of the glass?

Why It Works

A potato will balance over the point called the center of gravity. The long heavy forks help change the center of gravity to a lower point, making the potato more stable and easier to balance.

Balloon Blowout

Question

Are you strong enough to inflate a balloon inside a bottle?

Balloon Blowout

Materials

- one balloon
- one empty, narrow-necked glass bottle (vinegar bottle works well)

What to Do

1. Inflate your balloon. You may want to have a couple of extras in case yours pops. Was that an easy task to complete?

2. Hold up the glass bottle. What would it be like to first put a balloon inside the bottle (leaving the neck of the balloon on the outside of the bottle) and then try to inflate it? (Most responses: It would be easy, just like inflating the balloon in step one.)

3. After you have put a balloon inside the bottle, try to inflate it. What happens?

4. Why does your balloon not inflate inside the bottle?

Why It Works

There is already air inside the bottle when the deflated balloon is placed inside. When you initially blow into the balloon, it begins to inflate. This action traps the air already inside the bottle and creates a seal (when the inflating balloon begins to press against the bottle's neck). In order to inflate the balloon, compression of the air trapped between the balloon and bottle would have to take place. It is possible to compress the trapped air, but it would take great strength to do this. Human lungs are not strong enough to overpower the trapped air inside the bottle and compress it. The inside air pressure is, therefore, greater than the air pressure we are able to produce with our lungs, and the balloon cannot be inflated.

Bridge Construction

Does the construction of a bridge determine its overall strength?

Bridge Construction

Materials

- two pieces of paper
- three Styrofoam cups
- 12 pennies or washers

What to Do

1. Using two of the Styrofoam cups, place a sheet of paper on top of them, replicating a bridge.

2. Place the third cup on top of the bridge in the center. Observe what happens.

3. Repeat this experiment; this time, however, fold your paper like an accordion. Place this folded piece of paper on top of the two cups, replicating a bridge. Then place the third cup on top. Observe what happens.

4. Which paper bridge provides better support? Why?

5. Using your pennies or washers, one by one, place them into the cup on top of the folded paper. How much weight can it support? How many pennies or washers can you get in the cup? Why do you think the shape of the paper made the difference in the strength of the bridge?

Why It Works

When force or weight can be spread out or directed so that it is shared with many support structures, the form becomes stronger. The accordion fold spreads the force of the cup filled with pennies or washers, allowing more of the paper to be used in the support of the cup. Scientists know that the size and quantity of the folds are important. The larger and fewer number of folds, the stronger the paper. The smaller and greater number of folds, the weaker the paper.

Candle Caper

Question

Can you change the midpoint of a perfectly balanced candle?

Candle Caper

Materials

- one 8–10" (20–25 cm) candle
- craft knife
- two thin nails
- two soup cans (do not need to be empty)
- ruler
- sharpened pencil
- one 12" (30 cm) sheet of wax paper
- book of matches

What to Do

1. Write down what you think will happen when both ends of the candle are lit at the same time.

2. Using your ruler, measure the candle and find its midpoint. Mark the midpoint with a pencil.

3. Then push one nail into each side of the candle at its midpoint.

4. Place the two cans on a piece of wax paper, making sure there is room to balance the candle between them.

5. Put one nail on each can's rim, making certain that the candle is balanced.

6. Light the candle at both ends and observe it for a few minutes. What begins to happen?

7. What do you think makes the candle begin to seesaw?

Why It Works

When the unlit candle is resting on the two cans, the balancing point is in the middle. Once the candle has been lit, it begins dropping wax from each end. The balancing point moves to the opposite end, causing the candle to tip. Since both ends are dripping, the balancing point continuously moves back and forth between the two ends, causing a seesawing effect.

Clever Cup

Question

Will a cup burn just because there is a flame underneath it?

Clever Cup

Materials

- two 8-oz. (250 mL) paper cups
- one candle
- candle holder or modeling clay
- matches
- pitcher of water
- one large, empty wide-mouth glass jar with lid

What to Do

1. Place the candle in the candle holder or some modeling clay and then set it on a flat surface. What do you think will happen if you place the paper cup directly over the flame of the candle?

2. Fill the large glass jar with water and place it next to the candle holder. Then hold one of the empty paper cups over the candle flame and observe what happens. Be prepared to drop the cup in the jar of water once it catches on fire.

3. This time, fill the second paper cup 3/4 full with water. What do you think will happen this time when you hold the cup over the candle flame? Observe what happens.

4. Why did the water-filled cup not burn?

Why It Works

It takes four elements for fire to occur: (1) a heat source, (2) an oxygen source, (3) fuel, and (4) a kindling point. The first cup met all four criteria; the heat source was the candle, the oxygen source was the air around cup, the fuel was the paper cup itself, and the kindling point was the temperature at which the paper cup caught on fire. The second paper cup met only three of the needed criteria. The water in the cup acted as a heat conductor. A conductor allows passage of energy (i.e., an electric charge or heat). In this activity the water transfers the heat from the candle's flame through the paper cup and into the water itself. Therefore, the kindling point of the paper cup could not be reached, so no fire occurred.

Crafty Colors

Question

Can you make colors disappear?

Crafty Colors

Materials

- one 8.5" x 11" (22 cm x 28 cm) sheet of red transparent plastic wrap

- one 8.5" x 11" (22 cm x 28 cm) sheet of white paper

- one red and one blue crayon

What to Do

1. On a piece of white paper, draw a picture using your red and blue crayons.

2. Share your picture with a friend.

3. Using a piece of red transparent plastic wrap, cover the surface area of your picture. What happens when you do this?

4. Why do your red crayon color lines disappear when covered with red plastic wrap?

Why It Works

When light strikes a piece of paper it will absorb some of the rainbow colors and reflect others. The red crayon absorbs all of the colors and reflects only the red. The white paper absorbs none of the colors and reflects all of them. All the colors of the rainbow combine to form white.

The red paper acts as a filter. It allows some colors to pass through and others to be blocked. This makes it appear as if the red lines have disappeared. The red filter blocks out all of the colors from the white paper, allowing only red through. The paper and the letters are now both red appearing, making the red lines "disappear."

Cup-O-Strength

Question

Just how strong is the air pressure around us?

Cup-O-Strength

Materials

- one drinking glass or plastic cup

- one large index card

- water

What to Do

1. Fill the cup three-quarters of the way to the top with water.

2. Place an index card on top of the cup and rotate the card around the cup to make sure you have a complete seal.

3. Holding the card on top of the cup with one hand, slowly turn the cup upside down. Do not do this step too quickly.

4. Remove your hand from the card and observe what happens.

Why It Works

The air pressure pushing up on the card is much greater than the force of water pushing down. Air pushes with a force of approximately 14.7 pounds per square inch (6.6 kg^2). The water in this case (at sea level) has the pressure of approximately 0.14 pounds per square inch (.06 kg^2).

Dancing Raisins

Question

What causes raisins to rise and fall in a glass of soda?

Dancing Raisins

Materials

- a dozen raisins
- a clear plastic water container
- a one-liter bottle of soda water

What to Do

1. Fill your plastic water container with soda water.

2. Add the raisins one at a time to the soda water until all 12 are in the container. Watch as the raisins "dance" from the bottom to the top over and over again.

3. What is causing this to happen?

Why It Works

Soda water is carbonated with carbon dioxide gas, thus producing tiny bubbles when the cap of the bottle is removed. When you add raisins to the water, the carbon dioxide molecules fix themselves to the raisins' surfaces. This "focusing" of carbon dioxide molecules results in more buoyant raisins, allowing them to rise to the surface. Once the raisins reach the surface of the soda, the carbon dioxide gas is released, and they fall to the bottom of the container, where they repeat the process again.

Daring Dime

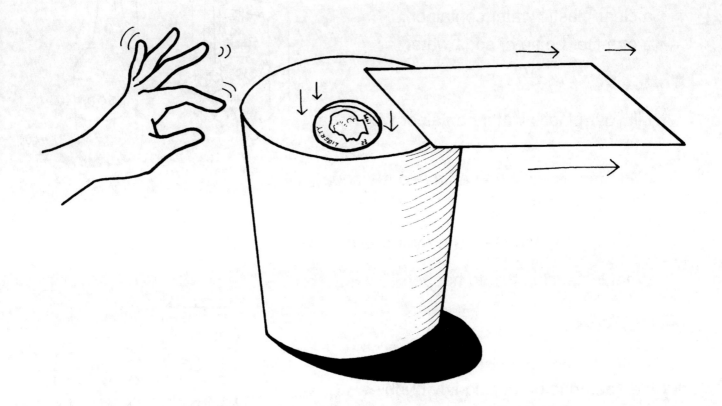

Question

What happens when you apply Newton's First Law of Motion?

Daring Dime

Materials

- one 6" x 2" (15 cm x 5 cm) strip of card stock paper

- an empty glass

- one dime

Note: Practice this activity several times before showing it to anyone.

What to Do

1. Center the card stock paper across the top of the glass.

2. Place a dime at the centerpoint of the paper.

3. Can you make the dime drop inside the glass without touching the glass or the dime?

4. Gather some of your friends and family around you as you perform this activity.

5. Using your thumb and forefinger, sharply flick the short edge of the paper in a forward direction. The card should fly forward, and the dime will drop into the glass.

6. Why did the dime drop into the glass and not fly forward like the paper did?

Why It Works

Newton's First Law of Motion states that a body at rest will stay at rest, and a body in motion will stay in motion. Both actions are based on inertia. To cause an object to start or stop moving, it needs to overcome inertia. The fast push you gave the card was enough to overcome the inertia of the card but not the coin. Since the card was moving forward and out from under the coin, the earth's gravity pulls the coin downward into the glass. Students can test the feeling of inertia for themselves the next time they ride in a car.

Dry Cleaning

Question

Does the air around us actually take up space?

Dry Cleaning

Materials

- one large clear, plastic or glass pitcher, 2/3 full of water
- two 8-oz. (250 mL) clear plastic cups
- one paper towel (cut in half)

What to Do

1. Crumple up each paper towel half and press the towels down inside the plastic cups so the paper towels rest snugly at the bottom. Turn both cups upside-down and shake gently to make sure the paper towels will stay in place.

2. What do you think will happen if you were to pour water into one of the cups?

3. Fill one cup with water and observe what happens.

4. What do you think will happen if you take the second cup and place it upside down into a pitcher of water?

5. Place the cup into the pitcher of water, open end first, and push it all the way to the bottom. Observe what happens.

6. Pull the cup back out of the water and feel the paper towel. Is it wet?

7. Why do you think the paper towel stayed dry while in the bottom of the water pitcher?

Why It Works

When the glass with the paper towel in it is put upside down in the pitcher of water, the air in the glass becomes trapped. The water that is in the pitcher pushes against the air at the same time the air pushes against the water, creating a barrier for the paper towel, keeping it dry. This proves that air does take up space.

Eggs-tra Strength

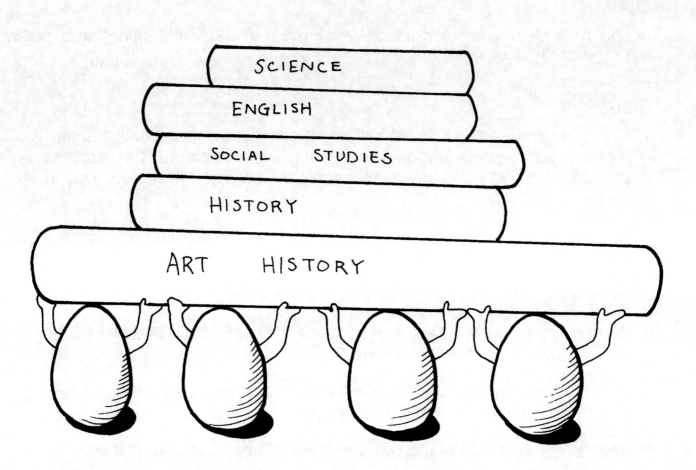

Question

How much weight can an eggshell or several eggshells withstand?

Eggs-tra Strength

Materials

- four raw eggs
- a craft knife
- one pair of scissors
- newspaper
- four paper towels
- several books

What to Do

1. Spread a few sheets of newspaper on a flat surface. Place two raw eggs in the center of the newspaper. What do you think will happen to the eggshells when you place a heavy book on top of them? After writing down some responses, place a book on the eggs, apply some pressure, and observe what happens. Compare the results to your responses.

2. Spread out some fresh newspaper. Place the remaining two eggs on the newspaper. Gently lay not one but three books on top of the two eggs. What do you think will happen?

3. Remove the books and tap the center of one egg sharply with your craft knife to produce a crack. Break the eggshell into two halves, placing the yolk in a bowl. Wipe out the inside of the eggshell halves with a paper towel. Repeat this process with the second egg. (If cracked edges are jagged, trim them using a pair of scissors, so they have an even edge.) Then place the egg halves on the newspaper so each half will be directly under one corner of the bottom book.

4. Gently place the first book on the egg halves. Then place the remaining books, one at a time, on top of the first book.

5. Observe what happens and answer this question: Why were the eggshells not crushed by the weight of the books?

Why It Works

The eggshells remained intact because of their dome shape. The weight on top of the eggshell dome is carried down along the curved walls to the wide base. No single point on the dome supports the weight of the object on top of it. The weight of the books is evenly distributed over the entire surface area of the domed eggshells.

Homemade Glue

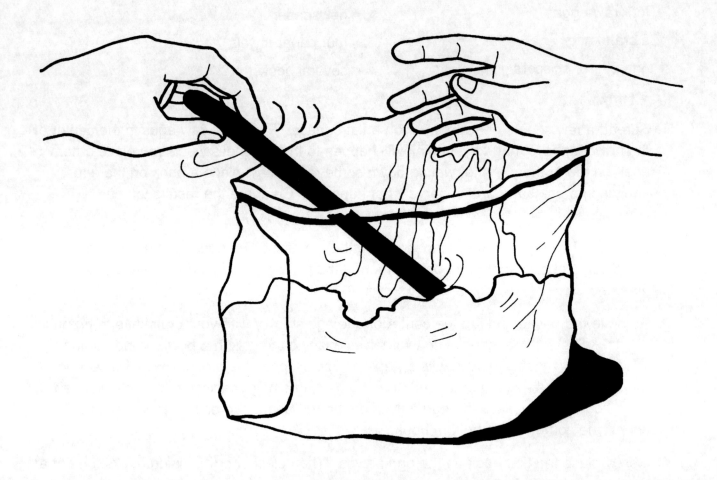

Question

How easy is it to make homemade glue?

Homemade Glue

Materials

- vinegar
- milk
- paper towel or coffee filter
- sealable plastic bag
- baking soda
- hot plate
- water

Note: Adult supervision is required for this activity.

What to Do

1. Stir .6-oz (20 mL) of vinegar into 4-oz (125 mL) of warm milk. To speed up the coagulation (to change a liquid into a soft, semi-solid, or solid mass), heat the mixture gently. (Do not boil.)

2. Filter the mixture (whey) through a paper towel or coffee filter and gently squeeze.

3. Place the curds from the towel or filter into a sealable plastic bag. Next add 1/4 teaspoon (1.25 g) of baking soda and one tablespoon (10–15 mL) of water.

4. Seal the plastic bag and knead the mixture until it is uniform. You have created homemade glue.

Why It Works

You have created a polymer. (A large molecule that is made by linking many small molecules together. Usually there are no more than three different types of molecules found in any one polymer.) The types of molecules help to determine its properties. In this case the molecules formed a glue substance.

How to Make Invisible Ink

Amy,
 We found the
treasure map
in the big

Question

Want to send a secret message? This is a great way for you and your friends to communicate back and forth, but only with this magical way can you send your secret codes.

How to Make Invisible Ink

Materials

- bottled lemon juice
- a thin paintbrush
- several pieces of notebook paper
- several long stick matches
- Note: Adult supervision is required for this activity.

What to Do

1. On a piece of blank notebook paper, you need to make up a secret message to send to a friend.

2. Using your paintbrush and lemon juice, paint your message on a second piece of blank notebook paper.

3. Place your paper message somewhere so the lemon juice will dry.

4. When your message has dried, have someone try to read it. When he/she gives up trying, light a match and sweep it in small circles about 1" (2.5 cm) away from the back of the paper. Be careful not to hold the match in one place for too long, or the paper may burn. After a few minutes, your secret message will appear.

Why It Works

Lemon juice is a chemical. When the small amount on the paper is heated, a chemical change takes place, allowing the secret message to be revealed.

I'm Crushed

Question

Can the change in air pressure affect something as insignificant as a plastic jug?

I'm Crushed

Materials

- a plastic gallon (4 L) jug with a screw-on lid
- a measuring cup with pouring spout
- hot plate (or other heating source)
- a pot
- a pot holder (optional)

What to Do

1. Pour one gallon (4 L) of water into your pot, place it on a heating source, and bring the water to a boil. While the water is heating up, fill the plastic with water (room temperature) until it is about three-quarters of the way full. Put the lid loosely on the jug, shake gently, and then screw the lid on tightly. Observe what is happening to the jug.

2. Empty the water from the jug and turn your attention to the boiling water. Carefully pour the boiling water into the measuring cup and then pour it into the jug. Repeat this process until the jug is about three-quarters of the way full. Put the lid loosely on the jug, shake gently, and then screw the lid on tightly. Observe what is happening to the jug this time.

3. Why do you think the jug collapsed after the boiling water was poured inside?

Why It Works

Equal air pressure exists inside and outside the jug when it is empty. The air pressure also remains equal when the jug is filled with tap water. The change occurs when boiling water is placed in the jug. The steam instantly begins to cool as it escapes from the jug (at this point, though, the escaping steam is not affecting the air pressure inside or outside the jug). When the cap is screwed tightly on the jug, the steam begins to condense, lowering the air pressure inside the jug. The air pressure outside the jug is now greater than the air pressure inside the jug, and it crushes the jug.

In Hot Water

Question

Why is hot water lighter than cold water?

In Hot Water

Materials

- two narrow-necked glass bottles
- hot water (needs to be very hot, but not boiling)
- one 4" x 4" (10 cm x 10 cm) heavy piece of cardboard
- blue food coloring
- cold water

Note: Adult supervision is required for this activity.

What to Do

1. Everyone knows hot air rises, but what about hot water? Predict whether you think hot water will rise like hot air.

2. Fill one glass bottle to its rim with cold water.

3. In the second glass bottle put 15–20 drops of blue food coloring, and then fill it with hot water. Cap the bottle momentarily and shake it so the food coloring can mix with the water. Remove the cap and make sure the bottle is filled with water.

4. Place the cardboard piece on top of the cold water bottle. While holding the cardboard in place, turn the bottle upside down and place the cold water bottle's rim directly on top of the hot water bottle's rim. Have a friend hold the bottom bottle steady.

5. While holding the top bottle steady with one hand, gently slide the cardboard out with your other hand. Suction will hold the two rims together.

6. Let go of both bottles and observe what happens.

7. What do you think causes the hot water to rise?

Why It Works

Hot air is less dense than cold air, allowing it to rise easier in the atmosphere. The same principle exists in water. The temperature of the water will determine how heavy it is. As water is heated, the molecules gain energy. They move faster, spreading apart, making the water lighter. The colder the water, the heavier it is. In the above activity, the hot water displaces the cold water, causing the cold water to descend into the hot water bottle; at the same time, the hot water rises into the cold water bottle until all of the water has been displaced.

It's in the News

Question

Can a newspaper break a ruler?

It's in the News

Materials

- one full-page (left and right sides) sheet of newspaper
- one thin meterstick or yardstick (without wire edges)
- a flat table

What to Do

1. Place the meterstick or yardstick on the table so that 12" (30 cm) of wood extends past the edge of the table. Ask a friend to assist you. What do you think will happen when your friend hits (with a direct downward motion) the portion of the meterstick or yardstick extending from the table?

2. Have your friend strike the meter or yard stick. Observe what happens.

3. Place the meterstick or yardstick back on the table with the same portion extending beyond the edge of the table. Open the newspaper so that the paper is one large sheet. Place the sheet over the meterstick or yardstick so the edge of the newspaper meets the edge of the table. What do you think will happen this time when your friend hits the meterstick or yardstick?

4. Have your friend strike the meterstick or yardstick, using the same downward motion as in step one. Observe what happens this time.

5. What do you think caused the wooden meterstick or yardstick to break?

Why It Works

Even though newspaper is very thin, it covers a large surface area. Air, which has a force of 14.7 pounds per square inch (6.6 kg^2), acts on the surface area of the newspaper. What all of this amounts to is approximately 9,000 pounds (4,050 kg) of air pressure pushing down on the newspaper. Because of this amazing amount of force acting on the newspaper, the wooden meterstick or yardstick that was underneath it could not move and actually snapped at its point of least resistance, where it began to extend from underneath the newspaper's surface.

Leap Frog? No... Leap Ball

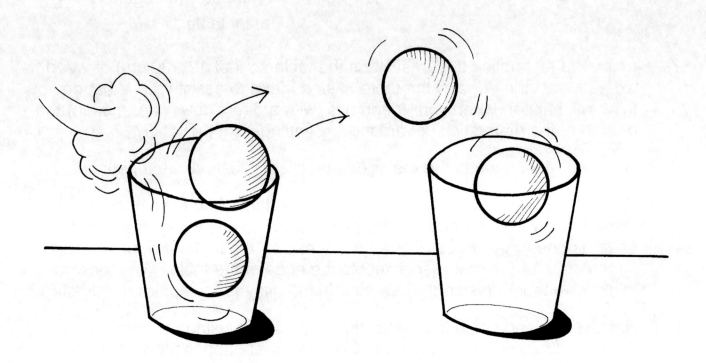

Question

Can you make a ping-pong ball jump from one cup to another without touching it?

Leap Frog? No... Leap Ball

Materials

- two 8-oz. (250 mL) wide-mouthed plastic cups

- a ping-pong ball

- a smooth surface

What to Do

1. Place the ping-pong ball in one of the cups and then place that cup on a smooth surface. Place the second cup next to first cup, leaving a one-inch (2.5 cm) gap between them. Brainstorm ways you could move the ball from one cup to the other. By the way, you cannot use your hands or any other props.

2. Now go ahead and try some of your ideas and see if they work.

3. Try this; blow into the far side of the cup containing the ping-pong ball with short, hard puffs. (Make certain you are blowing at an angle towards the far side.) After a few puffs, the ball will leap out and fall into the other cup.

4. What do you think caused the ping-pong ball to jump out and land in the second cup?

Why It Works

The ball leaped out of the cup due to an updraft of air pressure created by blowing into the cup at an angle. Because of the angle at which you blew into the cup, the air flowing over the ball formed an area of low air pressure, which caused the ball to move upward into the air in the room. The air current from the cup met the air current in the room and caused a low pressure flow, which, in turn, caused the ball to travel downward into the second cup.

Let It Rip

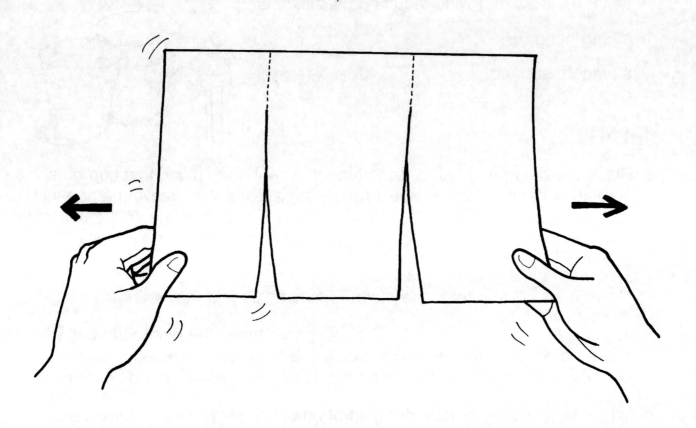

Question

What happens when you create force against your weakest point or points?

Let It Rip

Materials

- two 8.5" x 11" (22 cm x 28 cm) sheets of copy paper
- pair of scissors
- ruler
- pencil

What to Do

1. Fold one sheet of paper in half; crease the folded edge well. Open up the sheet of paper and, using your scissors, cut the crease line from the bottom to approximately 1" (2.5 cm) from the top edge. Then grasp the paper by the outer bottom corners and predict what will happen when you pull the paper apart. Now give it a try.

2. Take the second sheet of paper and fold it into thirds (you may want to use a pencil and ruler for an exact measurement). Crease the folded edges well. Then open up the paper and, using your scissors, cut each crease line from the bottom to approximately 1" (2.5 cm) from the top edge.

3. Again, grasp the outer bottom corners of the paper and predict what will happen when you pull the paper apart. After you have made your prediction, give the corners a pull and see what happens.

4. Why do you think the paper ripped into only two, rather than three, pieces?

Why It Works

A weak point is created by every cut made into a piece of paper. When force is applied (pulling apart) to a weak point in the paper, it will tear at that point. When two cuts are made in a piece of paper, you might naturally assume that there would be two weak points. This is not necessarily true. It is almost impossible to make two identical cuts with the naked eye. Therefore, the cut with the longer line becomes the weaker point. When you pull the paper apart, the force on the weaker point will cause the paper to rip there first.

Question

Can you make a marble move up the side of a jar without touching it?

Let's Go for a Spin

Materials

- one marble
- one large, narrow-mouthed glass jar (pickle jars work well)

What to Do

1. Place your marble on a flat surface. Do you think you can lift the marble off of the surface without touching it? Make some predictions as to what you think might happen.

2. Place the glass jar upside down over your marble so it is touching the inside curve of the jar. Hold onto the bottom of the jar and begin to rotate the jar rapidly. You can do it either clockwise or counter clockwise. Make sure you keep the mouth of the jar against the flat surface. Watch what happens.

3. Why do you think the marble climbed up the jar and stayed there as long as the jar was rotating?

Why It Works

When your marble begins to spin, force goes into motion. Centrifugal force is the force that causes an object to move away from a center of rotation. Newton's first law of motion states that an object at rest remains at rest and an object in motion stays in motion unless acted upon by an outside force. Centrifugal force results from the tendency of an object to continue moving in a straight line. If it is held back by another force, it cannot do that, thus the circular motion. The force that holds it back is called centripetal force, which is a force that causes an object to move towards a center of rotation.

Liquid Layers

← Alcohol

← Oil

← Water

← Glycerol

← Syrup

Question

Can you really mix oil and water?

Liquid Layers

Materials

- two tall, empty drinking glasses
- a measuring cup
- bottle of glycerol (available in most pharmacies)
- water (dyed blue with food coloring)
- bottle of alcohol (dyed green with food coloring)
- bottle of maple syrup
- bottle of cooking oil

What to Do

1. Place one empty glass on a flat surface. In random order, slowly pour 1/4 cup (63 mL) of each liquid into the glass. Observe what is happening in the glass. When you are finished observing, set the first glass aside.

2. Place the second glass on a flat surface. Slowly pour 1/4 cup (63 mL) of each liquid into the glass in the following order:

   First..........maple syrup

   Second.........glycerol

   Third..........blue-colored water

   Fourth.........cooking oil

   Fifth..........green-colored alcohol

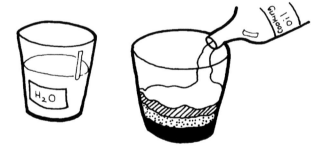

3. What caused all of the liquids to stay separated in the second glass?

Why It Works

All liquids have a specific density. Density is defined as the amount of weight an object (or liquid) has, given its size. By pouring the above liquids into a glass, using the specific order outlined in step three, the most dense (heaviest) liquid was on the bottom and the least dense (lightest) liquid was on top. The five layers remain separated because each liquid is actually floating on top of the more dense liquid directly beneath it.

Moving Grains of Pepper

Question

How can you make pepper run away?

Moving Grains of Pepper

Materials

- finely ground pepper

- small bowl or pie tin

- water

- liquid dish soap

What to Do

1. Fill the bowl or pie tin with water. It does not need to be full.

2. Sprinkle some pepper into the bowl or pie tin of water until you have an even layer of pepper floating on top of the water.

3. Carefully squeeze a drop of liquid dish soap into the middle of the bowl or pie tin. Observe what happens.

Why It Works

The surface tension of the water is what allows the pepper to float. This is a force which pulls the top of the water together and makes it act like skin. When a drop of dish soap is dissolved in the water, the surface tension becomes weak. The stronger forces in the tap water are then able to pull away from the weaker forces in the solution. The tap water gets pulled away from the soap and carries the pepper with it.

Musical Bottles

Question

How can you create music with a bottle?

Musical Bottles

Materials

- six empty, narrow-necked glass bottles
- water

What to Do

1. Fill each bottle with varying amounts of water. Start with one bottle being nearly empty to the last bottle being nearly full.

2. Once you have filled the bottles with water, line them up, one next to each other. You can place them either from the least to the greatest amount of water or from the greatest to the least amount of water.

3. Blow across the top of each bottle, one by one, starting with the bottle with the least amount of water. As you move from bottle to bottle, you will notice that each pitch gets deeper and deeper. Try it again, but this time, move as fast as you can across the row of bottles in one breath.

4. Once you become comfortable doing this, try to make a song with your musical bottles.

Why It Works

Sound is actually a wave, or vibration, in the air. The higher the pitch, the faster and shorter the waves or vibrations. The deeper the pitch, the slower and longer the waves or vibrations. Sound waves will vibrate in whatever space they have available. Blowing across the top of the bottles causes the air inside to vibrate. If waves have a large space to vibrate in, they will be longer and deeper, and if they have a small space to vibrate in, they will be shorter and sound higher.

Nifty Knives

Question

How can four knives hold up a ceramic bowl?

Nifty Knives

Materials

- four table knives (can be plastic)
- five ceramic bowls
  (soup or cereal bowls work well)

What to Do

1. Place four of the five bowls upside down on a flat surface. You are going to build a bridge using the four bowls and the four knives.

2. Using the illustration below as a guideline, interweave the knife blades, making sure that each handle is on top of one bowl.

3. When your bridge is complete, have someone flick the knives with their fingers at the side of the bridge's center. What happens?

4. Reweave the knives after the bridge has collapsed. Do you think this bridge can hold a fifth bowl after it collapsed so easily?

5. Place the bowl on top of your bridge. What happens?

6. How could the knives hold the weight of the bowl, when the bridge seemed so weak?

Why It Works

Each of the four knives is supported in two places—where the knife rests on top of the bowl (upward force) and where the knife is touching another knife (downward force). When the bowl is placed in the center of the bridge, the upward and downward forces become equal, cancelling the other one out (stress vs. strain) and friction takes over. The slight bend in the bridge when the bowl is placed on top stops when the added stress equals the strain. (Note: The degree of downward bend will vary depending on the weight of the object placed in the center of the bridge.)

One Tough Tissue

Question

Which is stronger, a broomstick or a facial tissue?

One Tough Tissue

Materials

- two facial tissues

- one rubber band

- one container of salt

- an empty paper towel cardboard tube

- one 7/8" (2 cm) wooden dowel (broomstick handle works well)

What to Do

1. Unfold one facial tissue and try tearing it apart. Why was it so easy to tear? In a moment you are going to try to tear the second piece of facial tissue, but first you will need to prepare it.

2. Unfold the second tissue and drape it over one end of your cardboard tube. Holding it firmly, attach it to that end of the tube with a rubber band. Pour approximately three inches (7.5 cm) of salt inside the tube, making sure the end with the tissue is resting on a flat surface. Tap the filled tube gently once or twice against the flat surface.

3. Have a friend assist you with this part of the activity. Hand your friend the tissue covered tube. Have him/her hold the tube while you push the dowel downward through the inside of the tube. What is happening? (Use caution with this activity. Close supervision is advised.)

4. What caused the facial tissue not to tear when the dowel was pressed down inside the tube?

Why It Works

Common sense would tell that the force being generated by the wooden dowel (broomstick handle) inside the cardboard tube would instantly tear the tissue. You need to consider the salt inside the cardboard tube and what its purpose is. Microscopic air pockets are in the layers of salt. When the dowel is pressed against the top layer of salt, it pushes the air out of the salt's air pockets and causes the salt to compact. The more compact the salt becomes, the higher its density. The salt can then absorb the force (shock) of the dowel's thrust and, in turn, causes only a small amount of the force being created to reach the facial tissue. Because of this reduction in force, the facial tissue does not rip.

Paper Porthole

Question

How is it possible for you to walk through an 8.5" x 11" (22 cm x 28 cm) piece of paper?

Paper Porthole

Materials

- several pieces of 8.5" x 11" (22 cm x 28 cm) paper

- pair of scissors

What to Do

1. Using a piece of paper and scissors, make a hole in the paper large enough to step through. You may not cut the paper up and then glue or tape it back together.

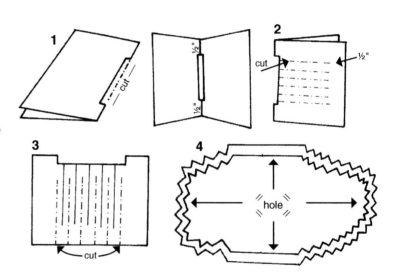

2. You may need several tries until you are able to figure this puzzle out.

3. Why do you think you are able to step through such a small piece of paper?

Why It Works

This activity replicates how a rubber molecule works. Rubber molecules are attached at their ends but not in the middle, allowing them to stretch and at the same time stay together. This is what you have created by cutting your paper and stepping through it.

Plunger Strength

Question

Why is it that when two rubber plungers are pushed together, it is difficult to pull them apart?

Plunger Strength

Materials

- two identical rubber sink plungers

What to Do

1. Wet the rim of each plunger, line them up facing each other, and push them together so they look like the figure below.

2. Ask a friend or a family member to help you pull the two plungers apart. You may find this very difficult to accomplish.

3. Try first breaking the seal of the two plungers and then pull them apart. Was it easier the second time?

Why It Works

When the plungers are pushed together, the air between them is squeezed out. When you stop pushing the plungers together, the rubber is allowed to return to its original shape. By squeezing out the air in the plungers, you have lowered the inner air pressure. The air pressure on the outside of the plungers is now greater, making it difficult to pull the plungers apart.

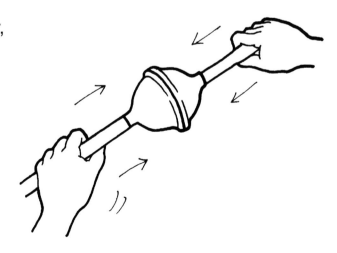

Powerful Paperbacks

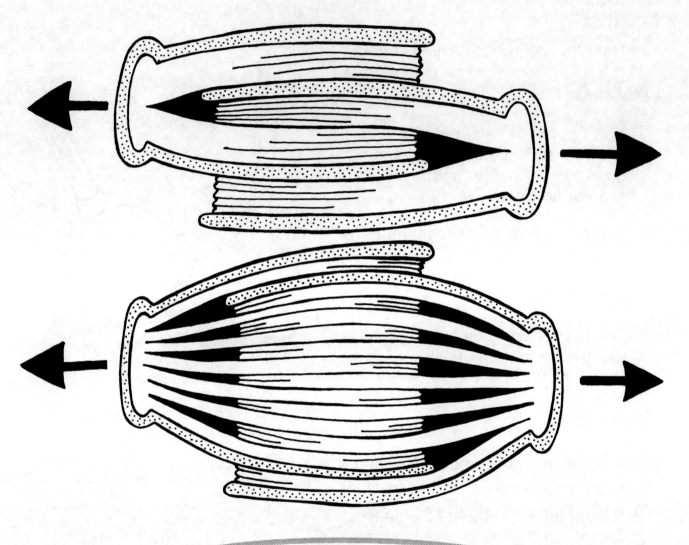

Question

Why do some books stick together like glue?

Powerful Paperbacks

Materials

- two paper or hardback books (same size and thickness)

What to Do

1. You will need a friend to assist you while doing this activity. Open each of your books at their centerpoints and interweave them. Then have your friend try to pull them apart. What happened?

2. This time take the same two books and interweave them again. This time use a shuffling effect. This should be similar to the way you would shuffle a deck of playing cards. (Note: Be certain to complete this process slowly so that book pages get as interwoven as possible.)

3. Hand the two interwoven books back to your friend and have him/her once again try to pull the books apart. What happened?

4. Why could your friend pull the books apart the first time, but not the second time?

Why It Works

The two books were unable to be pulled apart the second time due to friction. Friction can be defined as the resistance to motion of surfaces that are touching. Both of the books' pages are made of paper. If you were to look at one of the pages under a microscope, you would see that paper has a rough surface caused by the wood fibers in the paper pulp. When the pages are interwoven, the rough edges of the pages press hard against each other and will not slide apart. The harder you pull, the more friction you create.

Rocket Compression

Question

What will happen when you apply Newton's Third Law of Motion?

Rocket Compression

Materials

- one narrow plastic drinking straw
- one wide plastic drinking straw
- plastic squeeze bottle with screw-top lid
- thin cardboard
- glue
- modeling clay
- hammer and a nail

What to Do

1. Using the hammer and nail, make a hole in the screw-top lid just wide enough to slide your narrow straw through.

2. Slide your narrow straw through the hole in the lid, leaving about 4" (10 cm) of the straw sticking out at the top.

3. Seal the hole in the lid with a piece of modeling clay and then screw the lid tightly onto the plastic bottle.

4. Cut a 4" (10 cm) segment from the wide plastic straw and plug one end of it with modeling clay.

5. Then cut four identical triangles of cardboard and glue them to the sides of the wide straw at the bottom of the unplugged end. The triangles should form fins that stick straight out on four opposite sides of the straw.

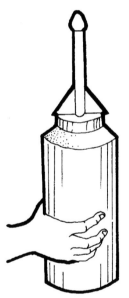

6. Finally, slide the wide straw over the narrow straw, give the plastic bottle a quick sharp squeeze, and watch your rocket soar.

Why It Works

When you squeeze the bottle, the air that is inside needs to escape. Because the wide straw is not permanently attached to the bottle, the trapped air is allowed to escape by pushing the wide straw up and away, freeing the air inside the bottle and causing the straw rocket to fly.

Sink or Swim

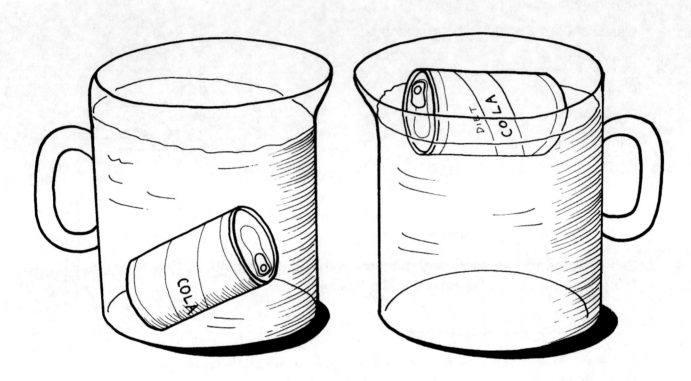

Question

Can a pitcher of water keep a can of soda afloat?

Sink or Swim

Materials

- one can diet soda (unopened)
- one can regular soda (unopened)
- a one-gallon (4 L) clear pitcher
- blue, red, and green permanent markers
- water

What to Do

1. Fill the pitcher two-thirds full with water. Mark the water line on the outside of the pitcher with a blue marker. Place the two cans of soda next to the pitcher. What do you think will happen when you place either can of unopened soda into the water?

2. Have a friend assist you with this activity. Have your friend place the can of regular soda into the pitcher of water. Observe the water level and mark it with a red marker. Did the regular soda sink? Remove the soda so the water level will return to its original blue marked level.

3. Do you think the results will be the same with the diet can of soda?

4. Have your friend place the diet soda can into the pitcher. Observe the water level and mark it with a green marker. Did the diet soda sink?

5. Why do you think the diet soda floated and the regular soda sank, maintaining the same water levels?

Why It Works

Liquids can have different densities. Density can be defined as the ratio of the mass of an object to its volume. Sugar is more dense than artificial sweetener. The regular can of soda contains sugar; the diet can of soda contains artificial sweetener. Both cans contain the same volume of liquid but do not have the same density of liquid. The difference in the liquid density between the two cans of soda is hard to detect by simply holding the cans in your hands. However, when they are placed in water, the different densities can be seen. They can also be observed by the difference in the marker lines. The lines indicate the amount of water displaced by the soda cans.

Spark in the Dark

Question

Can crunching on a piece of candy cause an artificial fireworks show?

Spark in the Dark

Materials

- two Wintergreen Lifesavers™ per person
- darkened room or outdoors at night

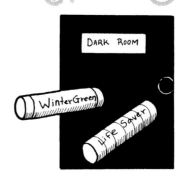

What to Do

1. You are going to make a mysterious light by crunching on special candy. You must have a friend for this activity.

2. Either in a room with the lights on or outside during the day, you will need to practice this procedure with your friend. The two of you need to face each other and should be no more than six inches (15 cm) apart. When a signal is given, you each put the candy in your mouth and begin crunching, while at the same time leaving your mouths open so you can see the other person crunch.

3. Now that you have practiced the crunching procedure, you are ready for the real thing. Take your second piece of candy and either go into a dark room or outdoors at night. Once your eyes have adjusted to the darkness, give the signal and crunch the candy with opened mouths while observing the candy being crunched in your friend's mouth. Observe what happens.

4. Why were sparks exploding while the candy was being crunched?

Why It Works

When wintergreen oil (a main ingredient in the candies) is crushed by the chewing motion and grinds with sugar (also in the candies) an electrical charge occurs. This charge is due to the fact that certain solids/liquids create electrical charges when fractured. The fracture (electrical charge) is visible as a light form (a small spark) and is commonly referred to as triboluminescence.

Star Search

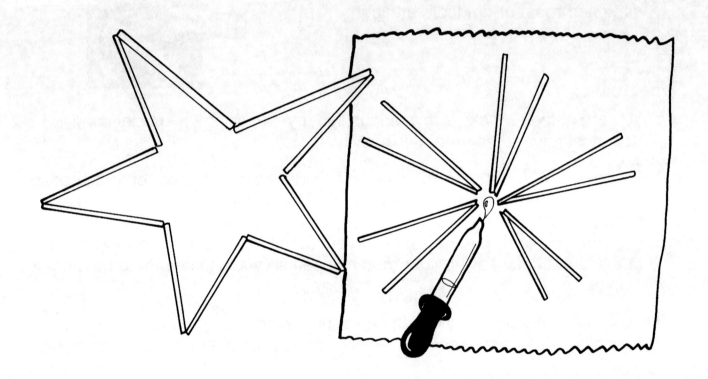

Question

What do stars and toothpicks have in common?

Star Search

Materials

- one cup of water
- a dozen flat-sided toothpicks
- an eye dropper, if available
- one sheet of wax paper
- a piece of paper
- one pencil

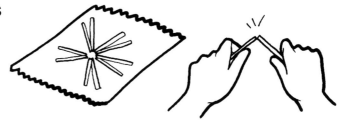

What to Do

1. With the help of a friend, place five flat-sided toothpicks on sheet of wax paper. Using them, create as many different designs as you possibly can. As you create your designs with the toothpicks, draw and name them on a blank sheet of paper.

2. Next you are going to match the design below by placing the toothpicks on the wax paper. As you do this, make sure each toothpick is broken in half right in the center, leaving the two halves together.

3. Predict what shape the toothpicks will make when you add one or two drops of water at the centerpoint of the broken toothpicks.

4. What caused the toothpicks to move and form a star?

Why It Works

Wood is made up of porous fibers. These fibers or capillaries act in the same manner as a sponge and easily absorb water. When the wooden toothpicks were snapped in half but still attached, their fibers also remained attached. When the water was dropped on the bent toothpicks, the fibers absorbed the water, causing them to expand. Because wood fibers have the natural tendency to lie straight, the water caused the bent portion of the toothpicks to move towards a straighter position, ultimately causing the star-like formation.

Sticky Penny

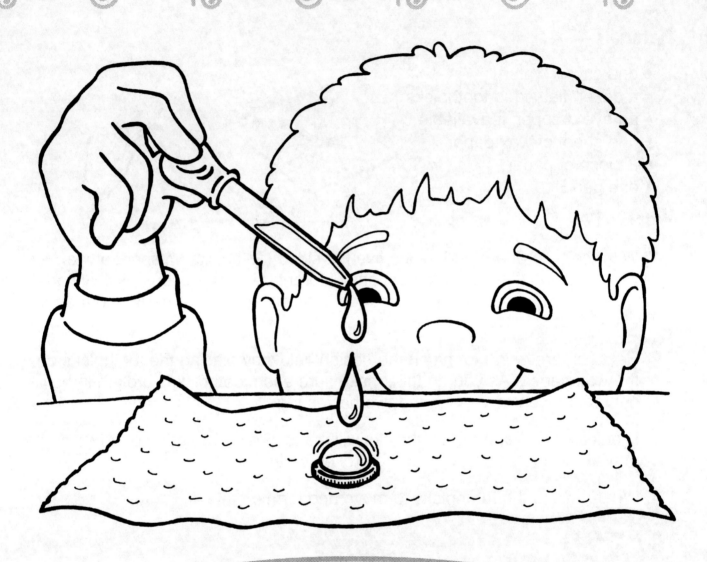

Question

Will water drops stick to a penny
and if so, how many?

Sticky Penny

Materials

- a penny
- eyedropper
- glass of water
- several paper towels

What to Do

1. On a table, place your penny on top of a paper towel.

2. Predict how many drops of water you can drip on a penny before the water spills over.

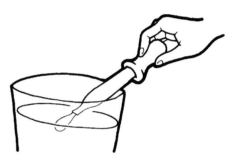

3. Fill your eyedropper with water and then begin dripping water onto your penny one drop at a time. Make sure you count your drops. Stop dripping water when it spills over the sides of your penny.

4. Repeat this experiment several times. Each time before repeating, be sure to dry off your penny.

5. Why does the water stay on top of your penny?

Why It Works

Cohesion, the magnetic attraction of similar molecules, causes water to cling together and mound up on certain types of surfaces. When this occurs, a skin-like surface forms on the water. This is called surface tension. When too much water is dripped on the penny, the surface tension will break, allowing the water to move freely. Once surface tension is broken, you cannot get it back. You need to repeat your experiment.

That's a Gong

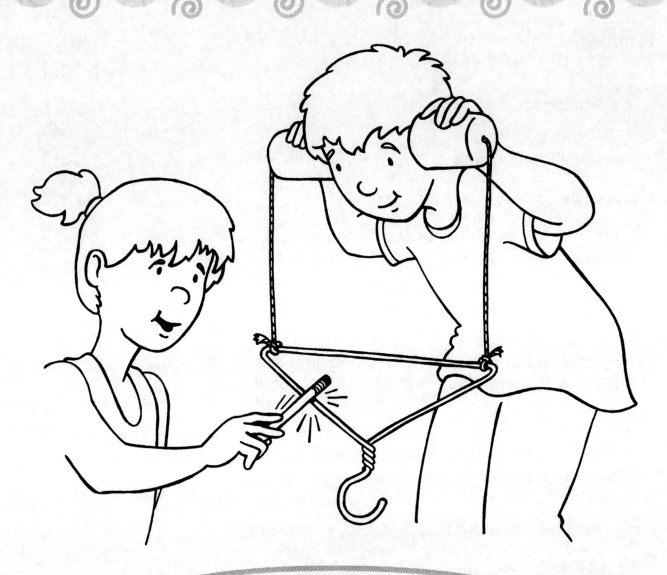

Question

If you struck several objects with the same hanger, would they all sound the same?

That's a Gong

Materials

- a wire coat hanger
- two pieces of string 24" (60 cm) long
- two small paper cups
- scissors

What to Do

1. What kind of sounds can you make with your fingers, using only one hand? Brainstorm as many as you can.

2. Using your scissors, poke a small hole in the bottom of each paper cup. Make sure it is big enough for the string to fit through.

3. Take one piece of string and thread it through the hole. Once it is threaded, tie a knot on the inside of the cup, big enough so the string will not come through when pulled. Repeat this step with the second cup.

4. When your cups are ready, tie the free ends of the strings to the opposite ends of the hanger.

5. Hold the cups up to your ears and hit the hanger against a solid object. Listen to the sound that is made.

6. Try hitting the hanger against several different types of objects to see if there are any differences.

Why It Works

Vibrations create sound waves. Sound waves travel though the air. We are unable to see them with the naked eye, but we can hear them. Because the air around us is so vast, sounds are actually hard to hear. When sound moves through solid objects (the hanger, string, and cups) it becomes more condensed. You hear it clearly because the hanger, string, and cups transfer the sound directly to your ear drums.

That's Egg-cellent

Question

What causes an egg that is bigger than the opening of the bottle it is sitting on to be pushed inside completely intact?

That's Egg-cellent

Materials

- one hard-boiled egg
- scrap paper
- matches or lighter
- an empty, narrow-mouthed glass bottle

Note: Adult supervision is required for this activity.

What to Do

1. Peel the shell off your egg and discard it. Then place the egg (narrow side down) on top of the mouth of the glass bottle. Why does the egg sit on top of the bottle and not fall in? What can you do to make the egg fall into the bottle without touching it? Brainstorm as many ideas as you can.

2. Lift the egg from the top of the bottle. Take a piece of scrap paper, light it on fire, drop it into the bottle, and then place the egg back on top with the narrow side facing down. Observe what happens.

3. What caused the egg to fall inside the bottle?

Why It Works

When you try to get the egg back out of the jar, the air pressure has again equalized, making it impossible. By blowing into the jar, you will increase the air pressure inside the jar, causing the egg to pop back out.

Initially, the egg sits on top of the glass jar because the air pressure inside the glass bottle equals the air pressure outside the glass bottle. When you place the burning paper inside the bottle, two opposing actions occur. The fire begins to heat the air inside the jar, making it more energetic. These fast moving molecules try to find a way out of the jar, but the egg resting on top is blocking their exit. The air continues to heat, and the pressure inside the jar is building up. The pressure becomes so great that it actually lifts the egg off the mouth of the jar. This releasing occurs several times, joggling the egg each time pressure is released. Another action that is taking place is the burning of oxygen. The oxygen changes from gas to solid by combining with the molecules of paper. Solid matter takes up much less space than gas, thus reducing the pressure in the jar. When the pressure is decreased enough, the egg is pushed into the jar because there is not enough air pressure on the inside of the jar to keep it out.

Watery Wonders

Question

Can you make an egg float in a glass of water?

82

Watery Wonders

Materials

- two large, clear drinking glasses
- two raw eggs
- container of salt
- a tablespoon
- warm tap water

What to Do

1. Fill one glass half full of warm tap water. Then add 10–12 heaping spoonfuls of salt, stir the mixture until the salt is completely dissolved, and set the glass aside.

2. Fill the second glass with water until the level equals the salt water level in the first glass.

3. Place the prepared glasses side by side on a flat surface. What do you think will happen when you put an egg into each glass?

4. Carefully place an egg in each of the two glasses. Why do you think one egg floats and one egg sinks?

5. Remove both eggs from both glasses and set them aside. Then empty half of the glass filled with salt water and slowly pour the plain tap water to refill the salt water glass. The two types of water will not mix. Gently place one egg into the glass. Observe what happens. The egg floats in the middle of the glass of water!

6. Think about why the eggs acted so differently in each glass of water.

Why It Works

Salt water is more dense (thicker) than tap water. Therefore, the egg which is less dense than the salt water is able to float on the top. On the other hand, the egg is more dense than the tap water, causing the egg to sink to the bottom of the glass. When the tap water is poured on top of the salt water, the tap water floats on the salt water. When an egg is added to this combination, it appears to be suspended in the middle, but it is actually floating on top of the salt water.

Wet or Dry

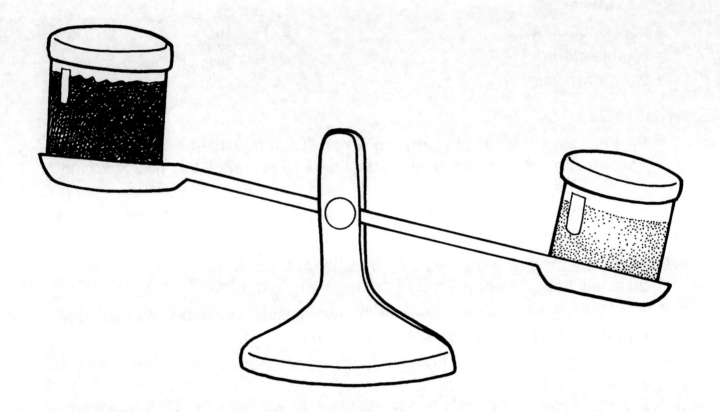

Question

Does sand have the same weight wet or dry?

Wet or Dry

Materials

- two wide-mouthed jars
- sand
- a pitcher of water
- a bowl
- a mixing spoon
- a measuring cup
- a weight scale

What to Do

1. Place both jars on a flat surface.

2. Using the measuring cup, fill jar one to just below the rim with dry sand. Record how many cups it took to fill the jar and set it aside.

3. In your bowl you will need to make wet sand. Slowly add water to the dry sand in the bowl and use the mixing spoon to stir. It is considered "wet sand" when it has the consistency of thick mud.

4. Then, using the measuring cup, fill jar two with the same amount of wet sand as dry sand used in jar one.

5. Weigh the two jars separately and record your findings.

6. Why does the dry sand weigh more than the wet sand?

Why It Works

The dry sand can fill all the spaces in the jar. It is very compact. When you mix sand with water, some of the spaces that were occupied by sand are now occupied by water, which weighs less. The sand cannot be compacted as much when water molecules are in the way.

What a Strong Grip

Question

Are you strong enough to pull a plastic bag out of a jar?

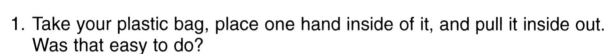
What a Strong Grip

Materials

- a wide-mouthed glass jar
- one resealable plastic sandwich bag
- one rubber band

What to Do

1. Take your plastic bag, place one hand inside of it, and pull it inside out. Was that easy to do?

2. Reverse the bag so it is no longer inside out. Do you think the bag would be just as easy to pull inside out if it were first placed inside a glass jar? Write down your prediction.

3. Place the bag inside the jar with the top of the bag folded over the rim of the jar. Then, secure it tightly with a rubber band below where the lid of the jar would be screwed on.

4. Now, try to pull the inside of the bag out of the jar. What happens?

5. What is stopping you from being able to lift the bag from the inside of the jar?

Why It Works

The bag cannot be pulled from the jar due to air pressure. When the bag is placed in the jar and the rim is tightly sealed, the air within the jar creates a vacuum. We are not strong enough to create an equal or greater force when trying to pull the bag out of the jar. The bag itself is also not strong enough to resist the force needed to pull it out from inside the jar. The bag would tear before being pulled out.

Wonder Boat

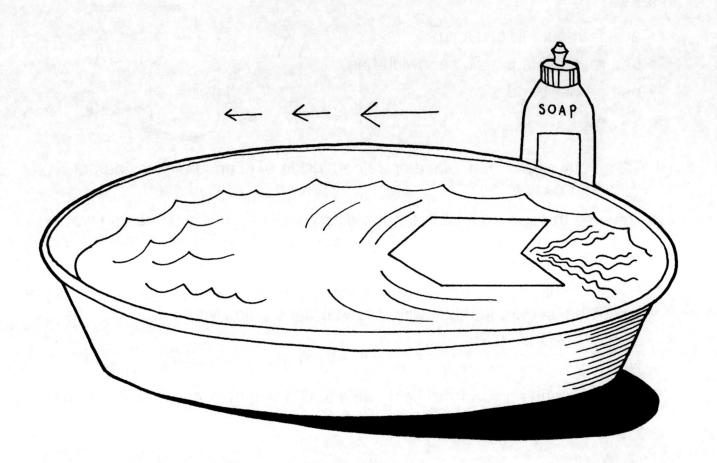

Question

Why does dish soap cause a boat to sail across a pan filled with water?

Wonder Boat

Materials

- boat pattern (page 90)
- pair of scissors
- a pie pan, filled with water
- a bottle of dish washing soap

What to Do

1. Using the boat pattern on the next page, cut along the solid black lines.

2. Next, cut your pattern in half, following the dashed lines, and set one half aside. With the other half, continue to cut along the dashed lines until you have what looks like a boat. Repeat this with the other half of your card.

3. Place the pie pan filled with water on a flat surface. This needs to be very still. Hold up one of your boats and have a friend observe your boat while it sits on the surface of the water.

4. Place your boat near the edge of the pan. What do your friend and you observe?

5. Remove the boat from the water and set it aside. Now, hold up the second boat and the dish washing soap. Will the second boat react any differently when you add soap to the water?

6. Place the second boat in the pie pan. It should be in the same location as the first boat. When the water is still, add one drop of dish washing soap to the water between the v-groove of your boat. Observe what happens.

7. Why did adding dish soap to the water cause the boat to move forward?

Why It Works

When dish washing soap is added to the surface of the water, the surface tension that exists relaxes. Therefore, when a drop of dish washing soap was placed in the water behind the v-groove of the boat, it broke the surface tension of the water, causing the boat to be pulled forward by the surface tension that existed in front of the boat.

Wonder Boat

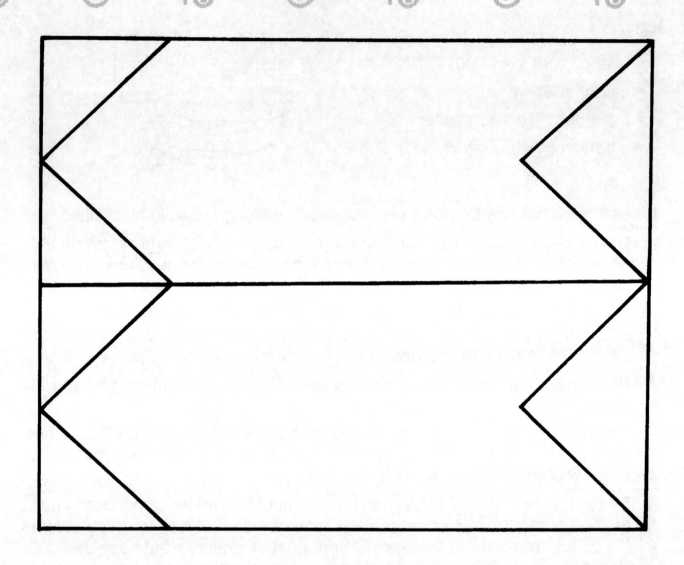

Life Science Activities

Animals Around the House

Question

Have you ever wondered what your family pet does during a typical day?

Animals Around the House

Materials

- pencil

- colored markers or crayons

- blank sheet of paper

- pet (can be the pet of a friend, neighbor, or relative)

What to Do

1. Choose an animal you would like to observe for this activity.

2. Set aside four time slots of at least an hour when you can watch your animal. Your observation can stretch over several days.

3. As you observe your animal, make sure you have paper, pencil, and markers or crayons handy to help you to record your observations. You will want to mark the date and time of each observation on your paper.

4. Now that you have had the opportunity to observe an animal for several hours, would you like to trade places with that animal, at least for a little while?

Ant Farming

Question

Do ants work together within their respective colonies?

Ant Farming

Materials

- a large, wide-mouthed glass jar
- a small glass jar that will fit inside the other glass jar
- cheesecloth
- strong rubber band
- moist soil, from outside
- fresh food scraps from various meals
- male and female ants, if possible

What to Do

1. Layer your large jar with about 2" (5 cm) of moist soil.

2. In your small jar collect several ants, tring to include a male and female, or you can order a colony from a science supply catalog.

3. Place your small jar of ants upside down inside your large glass jar.

4. Fill the excess space in your large glass jar with soil.

5. Add some fresh food scraps to the top layer of soil in your large jar.

6. Place a piece of cheesecloth over the lid of your jar and secure it with a large rubber band.

7. Observe your new ant farm and watch what they create. When you are finished, make sure you release your ants outside.

Why It Works

All ants are social. Ants that live in the same colony will work together to maintain the nest they live in. Each ant has one function, which will help its colony survive. Each of the three kinds of ants, queen ant, male ant, and worker ant, has a different job. Can you guess from the names what the jobs might be?

Checking Your Pulse

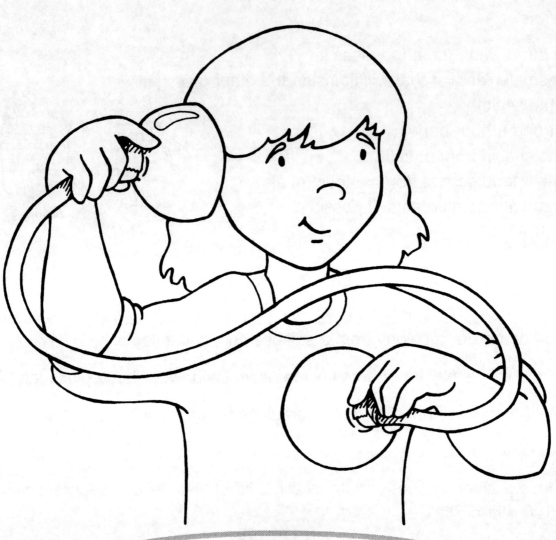

Question

Do you think that there is a direct connection between your pulse rate and your physical condition?

Checking Your Pulse

Materials

- 3' (.91 m) of rubber tubing

- two funnels
  (tops of plastic soda bottles work well)

- masking tape

- a watch with a second hand

- a blank piece of paper

What to Do

1. Insert one of the funnels into one end of the rubber tubing and attach it with masking tape. Insert the other funnel into the other end of the rubber tubing and attach it with masking tape. This is your homemade stethoscope.

2. While at rest, place one end of your stethoscope over your heart and the other end over your ear and listen. After a while, count the number of heart beats you feel in 15 seconds and then multiply that number by four. This is your resting heart rate.

3. Next, run in place for about one minute. Then, using your stethoscope, repeat step 2. This is your active heart rate.

4. Can you make a connection between your two heart rates and your physical condition?

Why It Works

Your physical condition and your activity determine how much energy your body needs. This energy requires oxygen which is supplied by your heart. When you are physically active, your body is using a lot of energy and a lot of oxygen. Your heart beats faster, and you breathe deeper in order to supply that oxygen.

Count The Insects

Question

Can you count the number of insects living within the boundaries of the hoop (string)?

Count the Insects

Materials

- large hoop or string
- paper and pencil
- magnifying lens (optional)

What to Do

1. Make a guess of how many insects live in your backyard or schoolyard. Be sure to write down your guess.

2. Brainstorm a list of ways you can use for finding the number of insects in an area.

3. Taking your hoop or string and paper and pencil, go into your backyard, park, schoolyard, or open field and place it down on the ground.

4. Carefully search the area that is within the hoop or string, counting all of the insects you see. Look in the grass, under sticks and leaves, on top of the dirt, on flowerheads, and in the air. Make sure to record the numbers and types of insects on your paper.

5. If you want to know the estimated number of insects in a particular area, you have to divide the area into square yards (m) with string, count the number of insects in a random selection of squares, find out the average, and multiply by the number of square sections. For example, if a field contained 15 square yards (12 m^2), a scientist would count the number of insects in four or five squares, find the average, and then multiply by 15 to get an estimated number of insects for that field.

Fishy Business

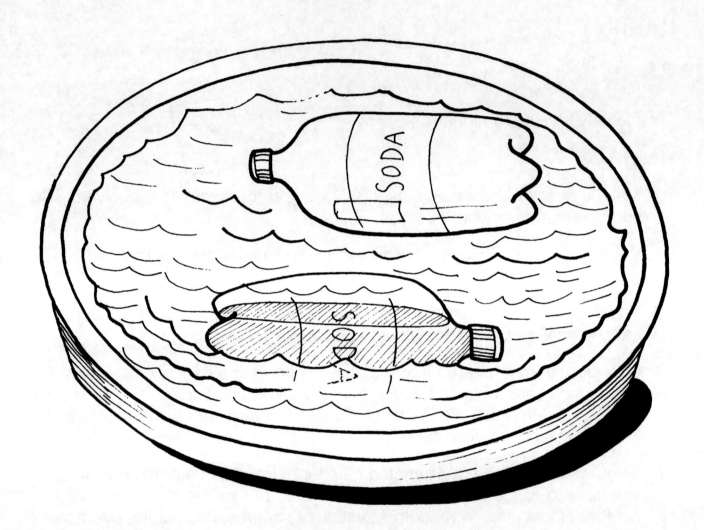

Question

How do fish make themselves float?

Fishy Business

Materials

- sink or plastic tub

- water

- plastic soda bottle, with its top

What to Do

1. Fill up your sink or tub with water.

2. Then, fill your soda bottle half way up with water and screw on the top.

3. Lay your bottle in the water and observe what happens.

4. Remove your bottle and fill it three-quarters of the way with water.

5. Lay your bottle back in the water and observe what happens this time.

6. Were the two outcomes different? Why?

Why It Works

Buoyancy is a force which causes objects to float in a liquid or to rise in air or gas. The less water you put in the bottle, the greater the amount of air pressure and the more buoyant the bottle. Fish control their buoyancy in much the same way. The amount of air they allow into their swim bladders will determine how buoyant they are.

Home For Winter

Question

How can you help a bird during the winter?

Home for Winter

Materials

- a 1/2 gallon (2 L) paper milk or juice container
- one yard (.91 m) of string
- scissors
- bird seed

What to Do

1. Cut out 2.5" x 3" (6.25 cm x 7.5 cm) windows on all four sides of the container.

2. Cut 2.5" (6.25 cm) slits down the sides of each window. Fold to form a perch for each window.

3. Poke six random holes in the bottom of the container for water drainage.

4. Punch two holes at the top of the container to attach string for hanging your bird feeder.

5. Fill the feeder with bird seed, hang it outside, and watch for the birds. It will not take them long to find your feeder.

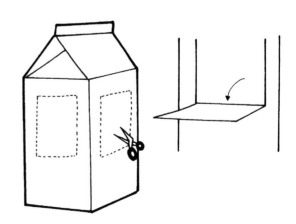

Ice Cream

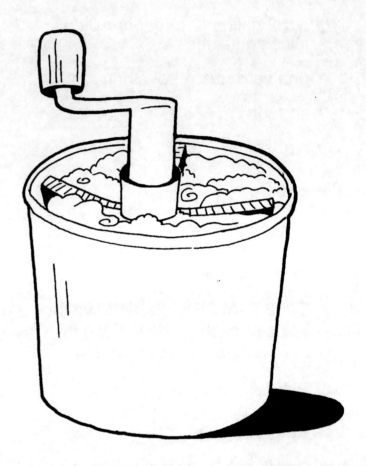

Question

What type of plant(s) might you find in store-bought ice cream that makes it different from homemade ice cream?

Ice Cream

Materials

- 3-pound (1.35 kg) coffee can and lid
- 1-pound (.45 kg) coffee can and lid
- masking or transparent tape
- wire whisk
- 1 cup (250 mL) whipping cream
- 1 cup (250 g) sugar
- 1 teaspoon (5 mL) vanilla
- 1 cup (250 mL) milk
- water
- crushed ice
- rock salt
- one pint (500 mL) of store-bought ice cream

What to Do

1. Mix in the 1-pound (.45 kg) coffee can all of the ingredients needed to make homemade ice cream (i.e., whipping cream, sugar, vanilla, and milk).

2. Fasten the lid tightly on the coffee can and tape it to prevent any leakage.

3. In the 3-pound (1.35 kg) coffee can, put a thin layer of ice and rock salt on the bottom of the can.

4. Place the 1-pound (.45 kg) coffee can inside the larger can and fill the extra space with ice and rock salt. Then fasten the lid tightly and tape it to prevent leakage.

5. Find an area on the floor where you and a friend can sit about 5' (about 2 m) apart. Roll the coffee can back and forth for 15 minutes, checking occasionally to see if the mixture in the smaller can has turned into ice cream.

6. When it has turned into ice cream, scoop it out into a bowl labeled "A." Then scoop out the store-bought ice cream into a bowl marked "B." Compare bowls A and B for differences and similarities in color, smoothness, taste, thickness, etc.

Why It Works

Store-bought ice cream contains a by-product of sea kelp called algin. Algin is used as an emulsifier and stabilizer in commercially made ice cream, as well as a variety of other foods. It works much like a preservative so that you can keep the food product longer.

Life in a Bottle

Question

Can plants live in a bottle despite the self-contained environment they are being placed into?

Life in a Bottle

Materials

- one 2-liter soda bottle
- scissors
- soil
- gravel or small rocks
- small plants found outside
- water

What to Do

1. Remove the label and cut the bottle in two about 1/4 of the way up from the bottom.

2. In the top portion, where it is cut away from the bottom, cut six vertical 1" (2.5 cm) slits spaced around the bottle.

3. Fill the plastic base of your bottle with gravel and small rocks.

4. Take the bottom of your bottle to a wooded area around your house. Dig up a few small plants, keeping the roots intact. Place the plants you collect (along with some soil from the same area) in the bottom of your bottle.

5. Water the soil and seal your plants by replacing the top portion of the bottle, fitting the slits down over the base.

6. Place it in a well-lit area and observe your terrarium for several days. If it is kept in a well-lit area, your terrarium will live for a long time.

Why It Works

Everything the plant needs to live (food, soil, and water) was placed inside the bottle before it was sealed. Then, when the bottle was sealed, a self-contained environment was created. Now, as long as the terrarium receives light, it will continue to live and grow.

Life Underground

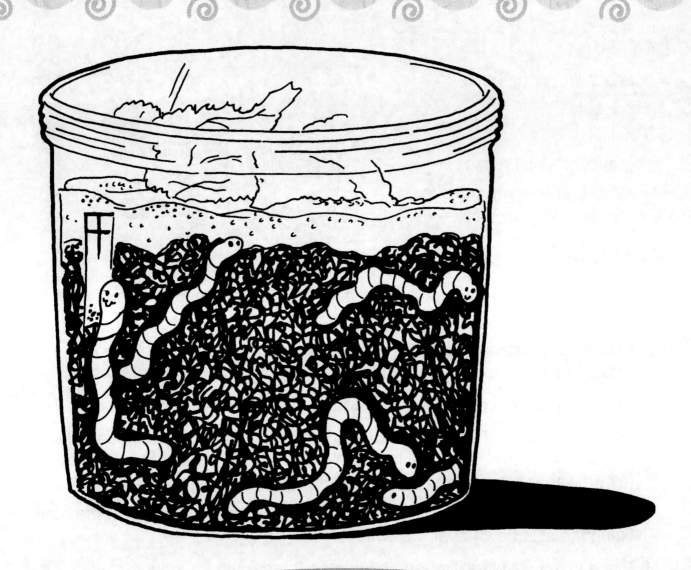

Question

Can you create a liveable habitat to support earthworms?

Life Underground

Materials

- a large, wide-mouth glass jar with a non-glass lid
- a hammer and nail
- an empty soup can with the lid removed
- soil
- sand
- 2 large sheets of black construction paper
- masking or transparent tape
- vegetable scraps
- earthworms

Note: Your teacher will need to use a hammer and nail to poke holes in the lid of the jar.

What to Do

1. Place the empty soup can, with the removed end facing down, in the middle of the glass jar.

2. Fill the glass jar with soil so it is level with the top of the tin can.

3. Add a thin layer of sand, approximately 1" (2.5 cm), to the top of the soil. Then throw in some vegetable scraps.

4. Add water to moisten the soil, being sure not to over water.

5. Cover the jar completely with black construction paper so no light gets into the jar.

6. Put a few earthworms in the jar, close the lid, and leave for it 24 hours.

7. Remove the lid and observe the changes inside the jar.

Why It Works

As long as you provide all of the necessary elements for a habitat (food, water, air, and space) you will have created a successful habitat for your earthworms.

Life's Little Building Blocks

Question

Have you ever wondered what plants, animals, and people are really made of?

Life's Little Building Blocks

Materials

- a small plastic bag
- one box of yellow gelatin
- several macaroni noodles
- several pieces of broken spaghetti
- a small ball (can be aluminum, clay, or a marble)
- water

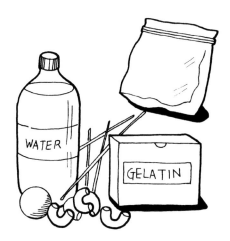

What to Do

1. Ask an adult to help you make the yellow gelatin. Add warm water to the gelatin mix to make it jelly-like. Allow it to cool for several minutes.

2. Then, pour half of the mixture into your plastic bag, add some of the macaroni and spaghetti, and let it begin to set.

3. Several minutes later, after the gelatin has begun to set, add the small ball, the rest of the noodles, and the remainder of the gelatin mixture. When all of your ingredients are in the bag, make sure to tie a knot in the top to keep everything inside.

4. You have created a replica of an animal cell. You may want to label the parts as follows: the small ball represents the nucleus, the macaroni represents the mitochondria, the spaghetti represents the endoplasmic reticulum, and the gelatin represents the cytoplasm.

Magic Plants

Question

Can some plants grow without seeds?

Magic Plants

Materials

- one potato
- one onion
- two glass jars, wide enough for an onion and potato to fit inside
- several toothpicks
- two containers filled with potting soil
- water
- a knife

What to Do

1. Fill both glass jars with water.

2. Insert toothpicks into each vegetable and suspend each in one of the glass jars. At least 1/4 of each vegetable should be covered in water.

3. Keep the jars filled with water.

4. When the shoots begin to grow and are at least 1" (2.5 cm) long, use a knife to cut out a chunk of each vegetable with the shoots still attached.

5. Plant each shoot in a container, cover it completely with potting soil, and then water.

6. Continue to water the containers each day and observe.

Why It Works

Plants are able to grow without seeds through a process called vegetative propagation. These plants regenerate themselves, using a part of the plant that already exists. Not all plants have this capability.

Moldy Bread

Question

What causes mold to grow?

Moldy Bread

Materials

- a loaf of sliced bread
- water
- plastic wrap
- magnifying lens
- a blank sheet of paper

What to Do

1. Take two slices of bread and moisten them both; wrap one in plastic wrap, leave one exposed to the air, and place them both on top of a counter.

2. Take two more slices of bread, placing one in the dark at room temperature and the other near a strong light source. They should both be exposed to the air.

3. Take two more slices of bread, keeping one dry and the other one wet. They should both be exposed to the air.

4. Take the final two slices of bread, placing one in a warm dark place and the other in a refrigerator. They should both be exposed to the air.

5. Examine each slice of bread once a day with your magnifying lens and record your observations.

Why It Works

Mold will develop on some of the above samples. In order for mold to grow well, it needs warm, dark, and moist conditions. Anything else will make it difficult for mold to grow.

P-U

Question

Do insects like everything they smell?

P-U

Materials

- several ants

- mint leaves
  (can be found at your local garden or grocery store)

What to Do

1. Go outside and find an area where there are ants.

2. Lay your mint leaves in a circle around several ants and observe for several minutes what happens.

3. Remove the mint leaves and again observe what happens.

4. Repeat this experiment with a variety of other scents (e.g., garlic, onion, cinnamon, etc.).

Why It Works

Just like people, insects do not like everything they smell. The ants stayed within the mint leaf circle because they do not like the smell of mint. When the leaves were removed, they were able to continue on their way.

Plants Eat Too

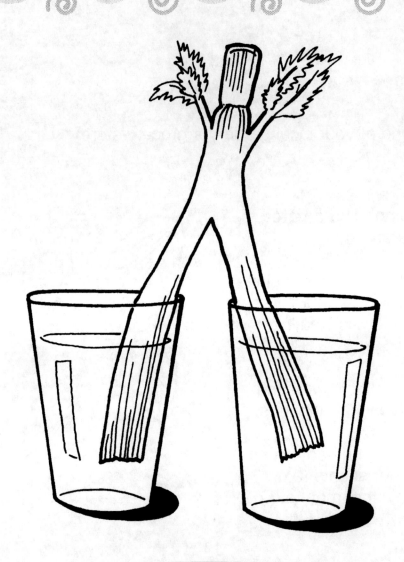

Question

Do plants use their feet in order to eat?

Plants Eat Too

Materials

- one white carnation or celery stalk
- two drinking cups (clear)
- food coloring (red or blue)
- scissors

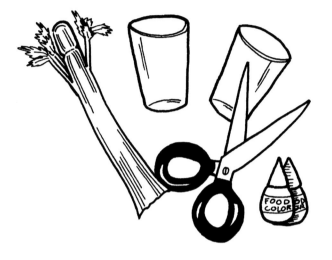

What to Do

1. Using your scissors, trim the stem of your carnation or celery stalk.

2. Fill your drinking glass 3/4 full of water.

3. Add several drops of blue or red food coloring until the color of the water has changed.

4. Place the carnation or celery stalk inside the glass and leave it there for several hours.

5. Observe your glass every 30 minutes and note any changes that are occurring.

Why It Works

Plants receive food and water through their root systems. This is made possible because of capillary action—the ability of water to flow against the pull of gravity by passing in and out of tiny plant cells packed closely together. The same process can be seen by dipping a paper towel in water.

Plant Parts

Question

How many different parts does a plant have, anyway?

Plant Parts

Materials

- paper lunch bag
- a large piece of construction paper
- white glue or transparent tape
- pencil
- colored markers or crayons

What to Do

1. Find an area outside or near your house where you have permission to collect plants and their parts (i.e., root, stem, leaf, flower, and seed).

2. Collect several different types of plants and their parts and carefully place them into your lunch bag.

3. When you return from collecting your plants, take your piece of construction paper and fold it in half.

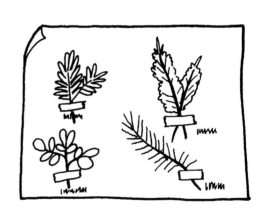

4. On one side, tape all of your collected plant parts and then label them. On the other side, draw a picture of each part you collected and also label each one.

5. Finally, design a picture that will go on the front of your construction paper to be used as the cover.

Run to the Sun

Question

Will a plant seek out sunlight in order to grow?

Run to the Sun

Materials

- 2 identical-sized shoe boxes with lids
- 8 cardboard strips that will fit inside the shoe boxes
- scissors
- 2 sprouting potatoes
- masking or transparent tape

What to Do

1. Cut a hole 2" (5 cm) in diameter at the end of one shoe box.

2. Using four cardboard strips in each box, create a maze and then tape the cardboard strips in place. If necessary, use your scissors to trim the cardboard strips so they fit inside the shoe boxes with their lids on.

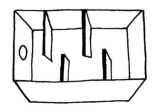

3. Place the sprouting potatoes at the ends of the shoe boxes. The potato in the box with the hole should be at the opposite end.

4. Cover both shoe boxes with their lids and place them on a sunny window ledge. Make sure the shoe box with the hole is facing the light.

5. Briefly remove the lids once each day for two weeks to observe the progress of the sprouting potatoes.

Why It Works

All plants need light to grow. Therefore, if light is available, a plant will search it out and grow towards it.

Slow Down

Question

What happens to insects in cold weather?

Slow Down

Materials

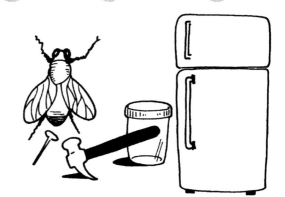

- a jar with a screw-on lid
- nail
- hammer
- a live insect
  (flies, ladybugs, or ants work well)
- refrigerator

What to Do

1. Using a hammer and nail, make several holes in the lid of your jar. Make sure you make them small so your insect cannot escape.

2. Place your live insect inside the jar and seal the lid.

3. Once your insect is inside the jar, observe it for about five minutes.

4. Place your jar inside a refrigerator for about five minutes. Then, remove the jar and observe your insect again for about five minutes. What are the differences?

Note: After your experiment is over, be sure to release your insect outside.

Why It Works

When you placed the jar inside the refrigerator, you reduced your insect's metabolic rate (i.e., the rate at which chemical changes take place in living things). As the metabolic rate of your insect slows, its movements also slow. As it returns to a normal temperature, so will its metabolic rate and its amount of activity.

Sprouting Sponge

Question

Is soil a necessary ingredient when you are trying to sprout seeds?

Sprouting Sponge

Materials

- a new dish sponge
- bean sprouts or mustard seeds (found at your local grocery store)
- a plastic plate
- plastic wrap
- a light source, preferably the sun
- a spray bottle
- scissors

What to Do

1. Decide on a shape for your sponge (triangle, star, heart, etc.). Then, using your scissors, cut it into that shape.

2. Soak your sponge in water, squeeze out any excess water so the sponge is still wet but not dripping, and then place it on a plate.

3. Evenly sprinkle your bean sprouts or mustard seeds over your sponge and then place your plate in a well-lit area.

4. Make sure to keep your sponge damp throughout the day by spraying it with a spray bottle. At night cover your sponge carefully with plastic wrap to keep in the moisture.

5. Continue to do this for two weeks, and by the end you will be able to taste what you have grown.

Why It Works

Seeds carry their own food supply, so when they are starting out, all they need is water, a light source, and the right temperature. Eventually, seeds that grow into plants need to be put into soil to continue growing.

The Same but Different

Question

What causes you to be similar to your parents and siblings but also different?

The Same but Different

Materials

- 8.5" x 11" (22 cm x 28 cm) piece of blank paper

- a pencil

What to Do

1. Turn your piece of paper horizontally and across the top write "Mom," "Dad," your name, and the names of your siblings (if you have any).

2. Beneath each name write down all of the physical characteristics you know about (e.g., eye color, hair color, hair type, etc.).

3. After you have written down all of the physical characteristics of your family, compare them to see what is the same and what is different from everyone else.

4. What do you think causes all of the differences and similarities?

Why It Works

When you look at yourself and then compare that to how your family members look, you are studying genetics. The way we look is decided by chromosomes. Chromosomes carry coded messages called genes that determine your hair color, eye color, etc. Your coded messages come from your parents, and that is why you look like them, as well as your siblings.

Tree Rubbings

Question

How can you make an artistic impression of your favorite tree?

Tree Rubbings

Materials

- pencil
- colored markers or crayons
- a large piece of construction paper
- glue or transparent tape
- scissors
- paper (tissue or tracing preferred)
- your favorite tree

What to Do

1. Take the piece of construction paper and fold it in half so it looks like a folder and then draw a picture on the cover, using the markers or crayons.

2. Go out into your neighborhood and find your favorite tree. Make sure you can reach some of the branches and leaves.

3. Observe your tree for awhile and write down the following information about your tree. How tall is your tree? How big is the trunk of your tree? What does your tree smell like? How old do you think your tree is?

4. Once you have finished writing down all of the above information, take the piece of paper and pencil and make a rubbing of the bark, a branch, and a leaf.

5. Then cut out each rubbing and glue or tape it inside your folder, labeling each tree part.

Underwater Locomotion

Question

How do fish and other finned aquatic animals move through the water?

Underwater Locomotion

Materials

- large balloon
- several small rubber bands
- stiff plastic (coffee can lid or milk container)
- large bowl or pan of water
- stapler or paper clips
- scissors
- plastic knife
- plastic drinking straw
- copy of page 134

What to Do

1. Fill your balloon up half way with water; do not knot it.

2. Over a sink or bowl, open your balloon enough to stick in a plastic drinking straw. Then hold the opening closed to keep water from shooting out.

3. Using a rubber band, attach the neck of your balloon to the straw, making sure it is tight enough so the water will not come out.

4. Bend the remaining part of the straw in half and attach a second rubber band so the straw will stay in place.

5. Cut the plastic into two fishtail-shaped pieces. Place each cut-out tail against the neck of your balloon and attach it with a rubber band. Then staple or paper clip the ends of the tails.

6. Put the plastic knife between the two pieces of the tail with the sharp end facing up. Then place the fish in the bowl or pan.

7. Lightly press your finger against the "nose" of the fish to keep it from moving from side to side. Then push the tail of your fish, using the plastic knife, and watch your fish swim.

Why It Works

Most fish gain forward movement and the ability to swim by using their tail fins. They do this by moving their tail fins from side to side. There are other fish that rely on the movements of their entire bodies to move through the water.

Underwater Locomotion

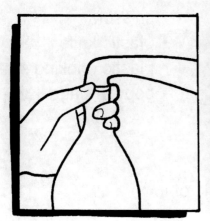
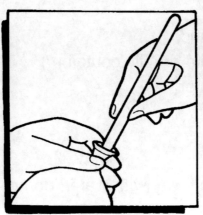
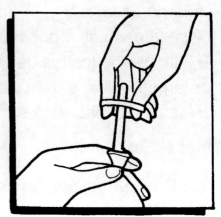
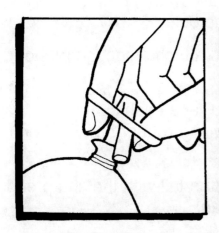

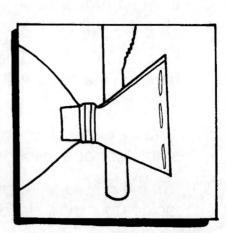

Earth
Science
Activities

Clean Water

Question

Is it possible to clean and filter dirty water at home?

Clean Water

Materials

- a 2-liter plastic soda bottle
- a large-mouthed glass jar
- scissors
- cotton
- pebbles
- gravel
- sand
- water
- a water pitcher
- soil
- a large spoon

What to Do

1. Using your scissors, cut off the bottom of your soda bottle.

2. Stuff a wad of cotton inside the top of your bottle and then turn it upside down so it is resting inside the glass jar.

3. Then, layer your bottle with the remaining materials. Each layer should be approximately 2" (5 cm). The order of the layers from the bottom up are pebbles, gravel, and sand.

4. Fill your pitcher with water and add two large spoonfuls of soil and mix.

5. Pour some of the dirty water into your filtering system and observe what happens.

Why It Works

The majority of water people use is not safe to drink without first filtering it. As you poured water into your filter, the impurities in it were held back by the different layers of filtering materials (i.e., sand, gravel, pebbles, and cotton). This process cleans the water by eliminating soil and organic matter but may leave behind harmful bacteria, making it undrinkable.

Cooking the Natural Way

Question

How can the sun help you to make a healthy snack?

Cooking the Natural Way

Materials

- one piece of tagboard 12" x 18" (30 cm x 45 cm)
- one piece of black construction paper
- aluminum foil (enough to cover tagboard)
- an apple
- plastic wrap

- scissors
- glue
- transparent tape
- several toothpicks
- large paper cup

What to Do

1. Glue a piece of aluminum foil as smoothly as possible to the tagboard. Use enough foil so the tagboard is completely covered.

2. Once the foil has dried to the tagboard, cut out a large circle.

3. Then, cut a circle out of the black construction paper that has a 2–3" (5–7.5 cm) diameter. Glue it to the center of your foil circle and allow some time for the glue to dry.

4. Once it is dry, cut a line from the outside of your foil circle to the center, shape it into a cone, and tape it in place.

5. Push a toothpick through the center of your cone. Attach an apple slice to the top of the toothpick so the apple is inside the cone and then cover the top of the cone with a piece of plastic wrap.

6. Place your cone inside a paper cup to stabilize it and then place your solar cooker in the sun to cook your apple.

7. When your apple has been cooked to your desire, enjoy a healthy snack, courtesy of the sun.

Why It Works

The foil reflects the sun's rays onto the black construction paper which then heats the apple. The apple is able to cook because the plastic wrap traps the sun's heat.

Expanding Water

Question

Can water and ice change the surface of the earth?

Expanding Water

Materials

- a raw egg
- drinking straw
- paper plate
- water
- small plastic cup to rest the egg on
- straight pin
- freezer

What to Do

1. Using a straight pin, poke a hole big enough to insert a drinking straw in the top of an egg.

2. Holding the egg over a paper plate, carefully blow into the straw, emptying the raw egg.

3. Fill your empty egg with water and stand it in the plastic cup so the water will not drip out.

4. Place the egg and cup in the freezer overnight.

5. Remove your egg and observe the changes that occurred.

Why It Works

Freezing the egg overnight simulates what happens to the earth's surface after years of exposure to water and ice. Water expands about 10% when it is frozen. This is enough of an expansion to crack your eggshell in much the same way ice can crack rocks on the surface of the earth. The plastic cup provides a catch basin for water that forms stalactites and stalagmites.

Homemade Volcano

Question

What causes the earth to erupt, releasing a fiery flow of lava?

Homemade Volcano

Materials

- 4 cups (1,000 g) of flour
- 1 cup (250 g) of salt
- 1 1/2 cups (375 mL) of warm water
- red food coloring
- water
- vinegar

- baking soda
- newspaper
- pie tin or plate
- large mixing bowl
- 3 8-oz. (250 mL) paper cups
- brown tempera paint (optional)
- paintbrush (optional)

What to Do

1. In a large mixing bowl, place 4 cups (1,000 g) of flour, 1 cup (250 g) of salt, 1 1/2 cups (375 mL) of warm water, and mix together, using your hands.

2. When your mixture becomes doughy, place it on a hard surface (kitchen counter, cutting board, etc.) and knead it until it becomes smooth and rubbery.

3. Mold your dough into a volcano, making sure to leave an opening at the top of the cone deep enough to conceal a paper cup. Bake your volcano in the oven for 30 minutes at 300° F (154° C).

4. Remove your volcano from the oven and allow it to cool. Place a paper cup inside the opening of your volcano. If you wish, you can paint your volcano brown and then allow time for it to dry.

5. Place your model volcano on top of some newspaper and prepare it for a simulated eruption.

6. Mix 3 tbs. (45 g) of baking soda with 4 oz. (125 mL) of water in a paper cup. Then add red food coloring to this mixture until it turns red and pour it into the cup inside the top of your volcano.

7. Pour 1 tsp. (5 mL) of vinegar into the cup inside your volcano and observe what happens.

Why It Works

When you combine baking soda and vinegar, a chemical reaction takes place. A gas is produced which causes your volcano to erupt, similar to the eruption of a real volcano. When pressure from heat, steam, and movement below the earth's surface reaches an extreme, lava will erupt from the top of a volcano.

Salty Water

Question

What is left after you evaporate ocean water?

Salty Water

Materials

- a sample of ocean water; if unavailable, make salt water

- a shallow pan

What to Do

1. Fill your pan with ocean water or salt water, depending upon what is available.

2. Allow the water in your pan to evaporate.

3. After the water has evaporated, taste what is left on the bottom of the pan.

4. What do you find in ocean water (salt water) after it evaporates?

Why It Works

Ocean water contains salt, which is a solid. When the water evaporates in the pan, the salt is left behind, leaving a light-colored residue on the bottom of the pan. The same thing happens with salt water.

Smooth Rocks

Question

Why are rocks that are found in rivers and streams smooth?

Smooth Rocks

Materials

- several small pieces of brick
- a plastic jar with a screw-on lid
- water
- magnifying lens

What to Do

1. Pick up one of the brick pieces and examine how sharp its edges are by feeling it. You may also want to examine it with a magnifying lens.

2. After examining the piece of brick, put all of the pieces inside your plastic jar, fill it with water, and seal the lid.

3. Taking turns with friends or family members, each of you should shake the jar about 100 times.

4. After everyone has had a turn shaking the jar, remove the pieces of brick and re-examine them for any differences. What has happened?

Why It Works

The swirling action of the water in the jar caused the rough surfaces of the bricks to wear down. If you were to continue shaking the bricks in the jar of water, they would keep getting smoother.

Soil Examination

Question

What do you think you will find if you dig in the soil?

Soil Examination

Materials

- 4' (120 cm) of string
- small trowel or shovel
- magnifying lens (optional)
- four craft sticks
- piece of paper
- pencil

What to Do

1. Using your string and craft sticks, go outside and find a place where you can measure off and mark 1 square foot (.09 m2) of soil.

2. Begin at one corner of your marked-off plot and carefully examine the soil.

3. If available, use a magnifying lens to better view things that are very small.

4. Draw a picture of each thing you see in the soil.

5. Label each picture after you have drawn it. If you are unsure of what you are drawing, use a reference book to identify your picture.

Soil: Plant, Mineral, or Both

Question

What makes soil?

Soil: Plant, Mineral, or Both

Materials

- rocks
- sand
- leaves and other plant material
- soil
- pie tin
- water
- magnifying lens

What to Do

1. Using your magnifying lens, examine a rock and then some sand. Look for their similarities and differences.

2. Originally, each grain of sand was part of a rock. Over time and with the help of erosion, the rocks were broken into smaller pieces, worn down, and mixed with other decaying materials, forming soil. This process takes a long time to actually happen.

3. In the bottom of your pie tin, spread a thick layer of sand.

4. Then, mix in ground-up leaves and plant material. You should use about the same amount of leaves and plants as you did sand.

5. Compare your mixture with the soil sample you already have.

6. If you have the time, moisten your mixture with water and set it aside for several weeks. Then compare your soil sample with your mixture one more time.

Why It Works

Soil is the combination of rocks and decaying organic matter that is mixed together and weathered over time. The above activity simulates this natural process but may not produce actual soil.

Stalactites/Stalagmites

Question

How do those magical icicles called stalactites and stalagmites really form?

Stalactites/Stalagmites

Materials

- 2 1-qt (1 L) glass jars
- a 24" (60 cm) length of soft yarn
- heating source
- a large cooking pot
- 8 oz. (226 g) of powdered alum (available at a local pharmacy)
- water
- two small rocks

Note: Adult supervision is required for this activity.

What to Do

1. In a large cooking pot, heat 2 qts. (2 L) of water. Do not bring it to a boil.

2. Pour 1 qt. (1 L) of the heated water into each of the two glass jars.

3. Dissolve 4 oz. (113 g) of alum into each glass jar.

4. Tie one small rock to each end of your yarn and then place the rocks inside the jars.

5. When the rocks have settled to the bottoms of both jars, set the jars in an area where they can be observed for several days without being touched. Make sure the two jars are close enough to cause a sag in the middle of the yarn.

Why It Works

As the water is soaked into the yarn, it will begin to form a small salt icicle which will hang down where there is a sag in the yarn. At the same time, the excess water in the yarn will drip from the sag onto the surface, and a small salt icicle will begin to rise. These are called stalactites and stalagmites which form because of the re-crystallization of the alum as the water evaporates.

Tasty Treats

Question

Can you make crystals out of sugar and water?

154

Tasty Treats

Materials

- medium-size sauce pan
- measuring cup
- four craft sticks
- stirring spoon (wooden)
- four small glasses
- sugar
- food coloring (optional)
- magnifying lens (optional)

Note: Adult supervision is required for this activity.

What to Do

1. In a medium sauce pan, combine 4 1\2 cups (1 L) of water and 2 cups (500 g) of sugar. The sugar may not dissolve immediately in the water.

2. Heat your solution until it comes to a boil, stirring occasionally. Continue to let it boil for two minutes.

3. Before pouring your solution into the four glasses, make sure the glasses are warm.

4. Pour a quarter of the solution into each glass, add food coloring, if desired, and stir with a craft stick.

5. Leave a craft stick in each of the four glasses.

6. Place the glasses in a well-lit area and let them stand for about a week.

7. Observe the glasses at least once a day while the crystals are developing.

8. At the end of the week, examine your crystals and taste what has developed.

Why It Works

When you heated the water, the sugar crystals dissolved, creating a supersaturated solution. As the water cools and evaporates, the sugar crystals will re-form along the craft sticks.

The Air We Breathe

Question

How polluted is the air that we breathe?

The Air We Breathe

Materials

- 3 index cards
- string
- transparent tape
- scissors
- magnifying lens (optional)
- black marker

What to Do

1. Using your scissors, cut a large hole in each one of your index cards.
2. Cover each hole in your index cards with transparent tape.
3. Again, using your scissors, poke a hole in each index card where you can tie a piece of string.
4. With a black marker, write "Please Do Not Touch" on each card so people will not disturb your test.
5. Choose three different locations where you can hang your index cards to collect possible pollutants.
6. Allow your cards to remain hanging for a week but check them daily for pollutant deposits.

Why It Works

Air pollution has become a daily part of our lives. Where one lives determines the amount of pollution found in the air we breathe. In this activity, particles floating through the air will affix themselves to the transparent tape on each of the index cards. After one week, any pollution that exists in your area will be collected on the tape. Many particles are very tiny and can best be seen through a magnifying lens.

Glossary

Air Pressure—the force that air puts on things. At the surface of the Earth, air pressure equals 14.7 pounds per square inch (6.6 kg/cm^2).

Algin—a by-product from sea kelp, used as a preservative in many different types of food.

Buoyancy—the force a liquid exerts on an object to keep it afloat.

Capillary Action—the ability of water to flow against the pull of gravity by passing in and out of tiny plant cells packed closely together.

Chemical—a substance that contains element(s).

Chromosomes—a threadlike strand of DNA that carries the genes and functions in the transmission of hereditary information.

Compression—the act or process or being pressed or squeezed together.

Condense—to change from a gas to a liquid.

Conductor—a material that allows heat or electricity to pass through it.

Contract—to get smaller.

Density—mass per unit volume. For solids it is mass in grams per cubic centimeter.

Dissolve—to break up and go into solution by the action of a liquid.

Element—a substance that cannot be broken down into a simpler substance.

Environment—all that is external to the living conditions of an organism.

Evaporation—the process of going from a liquid to a gas.

Glossary (cont.)

Expand—to get larger.

Force—a push or a pull that causes a change.

Friction—resistance to motion due to surfaces that touch.

Genetics—the study of heredity.

Gravity—the force that attracts objects towards the center of the Earth and keeps planets and other heavenly bodies in orbit.

Habitat—the environment in which a plant or animal can naturally be found.

Inertia—the tendency of all objects and matter to stay still if still, or, if moving, to go on moving in the same direction unless acted upon by some outside force.

Luminescence—an emission of light that does not derive energy from the temperature of the emitting body. It is caused by chemical, biochemical, or crystallographic changes.

Mass—the amount of matter usually expressed in grams (g) or milligrams (mg).

Matter—anything that takes up space.

Metabolism—the chemical and physical processes of maintaining life.

Metamorphosis—a change in the form and often habits of an animal during normal development after the embryonic stage.

Molecule—the tiniest particle of a compound.

Newton's First Law of Motion—a body at rest will remain at rest, and a body in motion will remain in motion at a constant velocity unless acted upon by an unbalanced force.

Glossary (cont.)

Newton's Second Law of Motion—force equals mass times acceleration.

Newton's Third Law of Motion—for every action there is an equal and opposite reaction.

Organic—pertaining to life.

Polymer—a long string of many molecules.

Predator—an animal that has to hunt down and catch its food in order to eat. Certain plants are also predators.

Prey—an animal that is killed for food.

Solution—a uniform mixture of two or more substances.

Stalactites—limestone deposits that grow downward from the ceilings of caves.

Stalagmites—limestone deposits that grow upwards from the floors of caves.

Surface Tension—the attractive force of the molecules of a liquid that pulls the surface tight to form an invisible film.

Vacuum—the absence of air molecules or other substances.

Vegetative Propagation—plants that are able to grow without seeds; instead, a part of the plant that already exists is used.

Velocity—a measure of a rate of motion in a particular direction.

Viscosity—a liquid's resistance to flow.

Volume—the amount of space something takes up. Mathematically, it is determined by multiplying together any item's width, length, and height.